30-SECOND
QUANTUM THEORY

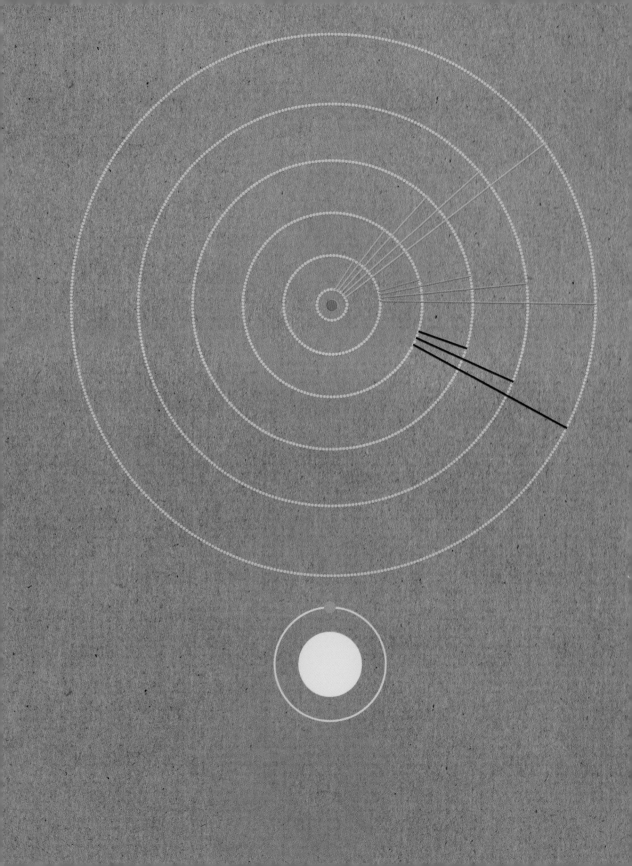

30-SECOND
QUANTUM THEORY

The 50 most thought-provoking
quantum concepts, each explained
in half a minute

Editor
Brian Clegg

Contributors
Philip Ball
Brian Clegg
Leon Clifford
Frank Close
Sophie Hebden
Alexander Hellemans
Sharon Ann Holgate
Andrew May

ICON

First published in the UK in 2014 by
Icon Books Ltd
Omnibus Business Centre
39–41 North Road, London N7 9DP
email: info@iconbooks.com
www.iconbooks.com

This book was conceived,
designed and produced by

Ivy Press
210 High Street, Lewes,
East Sussex BN7 2NS, U.K.
www.ivypress.co.uk

Creative Director **Peter Bridgewater**
Publisher **Susan Kelly**
Editorial Director **Caroline Earle**
Art Director **Michael Whitehead**
Project Editor **Jamie Pumfrey**
Designer **Ginny Zeal**
Illustrator **Ivan Hissey**
Glossaries Text **Brian Clegg**

ISBN: 978-184831-666-9

Printed and bound in China

Colour origination by
Ivy Press Reprographics

10 9 8 7 6 5 4 3 2

CONTENTS

INTRODUCTION
Brian Clegg

The physics we are taught at school can be,
frankly, rather boring. It is worthy, 19th-century science – necessary, certainly, but hardly earth-shattering. What a shame we don't introduce schoolchildren earlier to the more exciting bits: and there is nothing more thrilling and mind-boggling in all of science than quantum theory.

Small things
The idea that matter, stuff, is made up of tiny fragments goes all the way back to the ancient Greeks, but the ideas of the atomists (the name 'atom' comes from the greek word *atomos*, 'uncuttable') were largely sidelined by the four element theory that considered everything to be made up of earth, air, fire and water. By the end of the 19th century, atoms had made a comeback as useful concepts in chemistry and physics, but no one was quite sure what they were or how they worked. Some even doubted they existed. To the surprise of the scientists, atoms not only proved to be a reality, but these tiny components of everything from a human being to a speck of dust also behaved more strangely than anyone could expect.

It was initially assumed that atoms and their component parts would behave just like much smaller versions of the ordinary things that we see around us. Therefore, scientists thought that atoms would fly through the air just as a tennis ball does, if on a smaller scale. When it was discovered that atoms had internal structure, it was first theorized that they might be like plum puddings, with negative charges scattered through a positive body, but the revelation that most of their mass was in a central nucleus made it seem obvious that an atom was like a miniature solar system.

The quantum revolution
Unsettlingly for the old guard of physics (though delightfully for the rest of us), this picture turned out to be impossible to maintain. An atom built like a solar system would not be stable, and quantum particles refused to

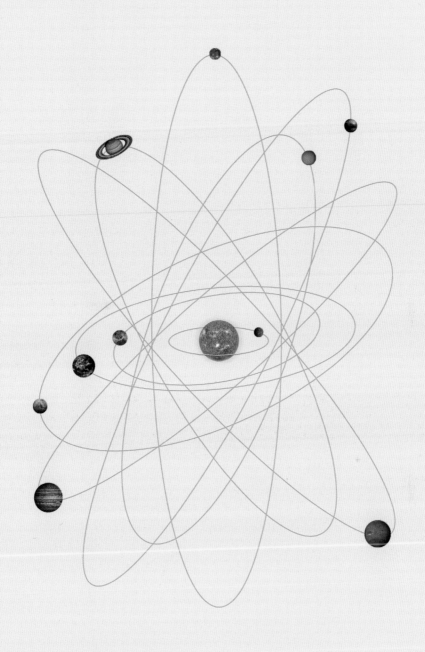

behave as predictably as a tennis ball. As quantum theory was developed it became clear that there was a fundamental difference between the microscopic and the macroscopic world. Tennis balls followed clear paths depending on their mass and the forces acting on them. But quantum particles could only be given probabilities of behaving in a particular way. At the heart of their behaviour was randomness and before they were observed it was never possible to be certain exactly what they were up to.

This horrified Einstein, inspiring him to write 'I find the idea quite intolerable that an electron exposed to radiation should choose of its own free will, not only its moment to jump off, but also its direction. In that case, I would rather be a cobbler, or even an employee in a gaming house, than a physicist,' and produced his famous comments along the lines of 'God does not play dice.' But others were fascinated.

The great American quantum physicist Richard Feynman said 'I'm going to describe to you how Nature is – and if you don't like it, that's going to get in the way of your understanding it... [Quantum theory] describes Nature as absurd from the point of view of common sense. And it agrees fully with experiment. So I hope you can accept Nature as she is – absurd.' I hope that this book will give you a chance to experience and echo Feynman's enjoyment of the sheer strangeness and absurdity of the quantum world.

Dividing the quantum

There surely can be few topics that are more suited to being divided up in bite-sized, easily digested chunks than quantum theory. The 50 articles that follow are split into seven sections, each taking on a part of this significant undertaking. We begin, aptly enough, with **The Birth of the Theory**, which describes how the classical view that atoms were no different from the objects we see around us was undermined by observation and how the endeavour to find a way that atoms could be stable at all required a very different approach.

From the origins we move on to **The Essentials**, the key components of quantum theory, some of which, like Heisenberg's Uncertainty Principle, have escaped the confines of physics to become part of popular culture. With these fundamentals in place we can open up the science of practically all of our everyday experience, **The Physics of Light & Matter**. Quantum electrodynamics, the theory that explains everything from the way sunlight can warm us to the reason why you don't fall straight through your chair, required a whole new way of looking at the quantum world and has become the most successful theory ever in terms of the accuracy with which it predicts what is observed.

From here we move onto some key **Quantum Effects & Interpretation**, explaining how we can both see through and see a reflection in a window, or how quantum tunnelling keeps the Sun working – and the thorny matter of quantum interpretation. Quantum theory is almost unique in this respect. It is excellent at predicting what we observe, but no one is sure exactly what the theory itself represents. Descriptions like the Copenhagen, Many Worlds and Bohm interpretations attempt to put what is observed into a framework that explains why such observations are made, yet we have no way to distinguish between these options, choosing on personal preference rather than good scientific logic.

In the next section we encounter quantum theory's most remarkable phenomenon, **Quantum Entanglement**. Described by Einstein (who hoped to use the concept to disprove quantum theory) as 'spooky action at a distance', entanglement makes it possible for one quantum particle to influence another instantly at any distance, seemingly in contradiction to special relativity's limit of the speed of light – and yet experiment after experiment has confirmed its existence, and applications like quantum encryption and quantum computers rely on entanglement to work.

The final two sections look at the ways in which technology based on quantum theory has penetrated our everyday lives, and the extreme possibilities that quantum physics makes possible. In

Quantum Applications, we discover the laser, the transistor, the MRI scanner and more. Whenever we use electricity we are making use of a quantum phenomenon, but electronics has made an explicit knowledge of quantum theory an important part of the design of technology, to the extent that it is estimated that around one-third of the GDP of developed nations comes from technology based on quantum theory.

As for those **Quantum Extremes**, here we can take in the mystery of zero point energy, beloved of fringe science, but a real quantum effect that means that even a vacuum is not empty, the peculiar behaviour of extreme low temperatures, and the extension of quantum theory into the atomic nucleus, gravity and even biology.

Jumping in

Each topic, supported by a exciting illustration, is broken down to make it accessible. The 30-second theory section gives the main description, while the 3-second flash summarizes the topic at a glance. If you would like to find out more, the 3-minute thought takes a particularly intriguing aspect of the topic and expands on it. The related theories point to other topics that will follow on naturally, while the 3-second biographies identify key names in the development of this area.

The format of 30-Second Quantum Theory is itself quantized, breaking up the essentials to discover, enjoy and absorb what is arguably the most fascinating and mind-bending aspect of all science. Everything we do, everything we see, has quantum particles at the heart of the action – and yet these particles are so very different from anything we ever directly experience. That's the paradox and the delight of quantum theory, as you are about to discover.

THE BIRTH OF THEORY

black body A hypothetical object absorbing all light that hits it, whatever the frequency or direction. A black body at a constant temperature emits a light spectrum (black body radiation) purely dependent on its temperature and not influenced by the nature of the body.

black holes A location at which matter has been made so compact that it collapses to a point under gravitational pull. Most frequently formed by the collapse of a massive star. The apparent size of the black hole is its 'event horizon', which is the distance from the centre where nothing, not even light, can escape. The black hole itself is a singularity, a dimensionless point.

complementarity Because in quantum theory the act of measurement influences the result, different measurements are complementary to each other. So, for instance, depending how you make a measurement on light it can appear to be a wave or a particle, but not both at the same time. Complementarity states that reality is neither of these but a whole of which we can only detect a part with any one experiment.

frequency The number of times a repeating phenomenon occurs in a second. Often used for a wave, describing the number of cycles the wave makes in a second (measured in hertz, where 1 Hz is one cycle per second). A wave's frequency is its velocity divided by its wavelength. For a quantum object the frequency is proportional to the energy of the object.

Hawking radiation This quantum effect, predicted by Stephen Hawking, is produced when virtual particles briefly appear and disappear in space. Usually they leave no trace, but if this happens near a black hole's event horizon, one particle can be pulled into the hole while the other flies off, creating radiation. (So black holes aren't truly black.) Hawking radiation is an example of black body radiation, equivalent to that of a black body at a temperature inversely proportional to the black hole's mass.

lepton A fundamental particle with a quantum spin value of 1/2, the best known example of which is the electron. Other leptons are the muon, the tau and the three types of neutrino.

photon A quantum particle of light and the carrier of the electromagnetic force. Until the 20th century light was thought to be a wave, but both theory and experiment show that it is also a massless particle.

Planck's constant A fundamental constant of nature, technically the 'quantum of action', where action is a mathematical representation of a system's energy as it moves along a path. The constant, denoted by Planck himself as h, describes the relationship of the energy of a photon to its frequency (colour). It is very small: just over 6.6×10^{-34} joule seconds.

Planck's formula/relation The relationship between the energy of a photon and its frequency, given by $E=h\upsilon$ where h is Planck's constant and υ is the frequency.

quantum leap Despite its popular use to describe a significant change, a quantum leap is usually a tiny jump – the change between two levels of a quantized system, for instance the jump an electron makes between adjacent electron orbits.

quanta The plural of 'quantum' (literally 'how much', as in quantity), but used to indicate a particle or 'packet' providing the minimum unit of energy or matter. Hence 'quantum theory', describing the behaviour of particles of matter and of light. When a value is 'quantized' it comes in discrete amounts. An average family, for instance, may have 2.3 children, but children are quantized, so an actual family can only have a whole number of children.

wavelength The length of a repeating wave in which it passes through a complete cycle, returning to the starting point of the cycle. Wavelength is velocity divided by frequency.

THE ULTRAVIOLET CATASTROPHE

the 30-second theory

Popular histories of quantum mechanics generally award a central role to the so-called ultraviolet catastrophe. Physicists at the end of the 19th century realized classical physics predicts that electromagnetic energy radiated by a 'black body' – which we can envisage as a warm, totally light-absorbing object – becomes infinite for wavelengths shorter than those of visible light, in the ultraviolet part of the spectrum. This was obviously wrong, and in 1900 Max Planck found he could avoid the 'catastrophe' by assuming that the oscillating atoms in the black body could only emit energy in discrete packets – quanta – of a size proportional to their vibration frequency. He denoted the constant of proportionality as h, now called Planck's constant. Exactly how Planck viewed his 'quanta' is still disputed, but he seems to have resisted for years the idea that they corresponded to reality, seeing them instead as a mathematical convenience. That they resolved the ultraviolet catastrophe – by imposing restrictions on the ways a black-body atom can oscillate at high frequency and thereby reducing the energy they radiate – was only pointed out later. Still, the idea won Planck a Nobel prize in 1918.

RELATED THEORIES
See also
PLANCK'S QUANTA
page 18

EINSTEIN EXPLAINS THE PHOTOELECTRIC EFFECT
page 20

3-SECOND BIOGRAPHIES
MAX PLANCK
1858–1947
Regarded in the early 20th century as the elder statesman of German physics

WILHELM WIEN
1864–1928
German physicist who found experimentally how the intensity of black-body radiation at different wavelengths depends on its temperature

30-SECOND TEXT
Philip Ball

3-SECOND FLASH
The hypothesis proposed by Planck that energy is quantized (divided into chunks) avoided the ultraviolet catastrophe predicted by classical physics for black-body radiation.

3-MINUTE THOUGHT
A perfect black body seems an oddly exotic entity, but it is really just an idealization of a warm body. The basic idea that the hotter the body is, the more short-wavelength radiation it emits, is familiar from the glow of an electric bar heater as it warms, and applies to stars too. Even black holes emit black-body radiation, in the form of so-called Hawking radiation. The black hole acts rather like a black body with a temperature inversely proportional to its mass.

Unless light is quantized, a black body should emit radiation uncontrollably.

PLANCK'S QUANTA

the 30-second theory

3-SECOND FLASH
Planck discovered the quantization of energy – the fact that energy is given off or absorbed by matter in discrete packets or quanta – which revolutionized physics.

3-MINUTE THOUGHT
Initially Planck viewed the quantum as a mathematical construct. Physicists paid little attention to it until Einstein, in 1905, explained the photoelectric effect by equating Planck's quanta with photons. The real physical nature of quanta subsequently became clear, when Bohr explained how electrons can only move in fixed orbits around atomic nuclei, and every time they change orbit they do so by giving off or absorbing a photon.

In the late 1890s a German manufacturer of light bulbs asked a young German physicist, Max Planck, to compute the energy emitted by the hot filaments in light bulbs. Here Planck was faced with a problem that physicists at the time could not solve: to find a formula for the distribution of wavelengths of light and infrared light given off by a black body at any given temperature. Planck tried everything physics theory could offer, to no avail. In what he called an 'act of desperation' he introduced the notion that the radiation from a hot body is not given off as a continuous stream, like water running from a tap, but rather like a dripping tap, in tiny packets – quanta – which he initially called 'energy elements'. He assumed that the energy of these packets was inversely proportional to their wavelength, the energy of the packets with the shortest wavelengths being the highest. The relation between the wavelength of these quanta and their energy became known as the Planck relation. When Planck applied this idea to the computation of the energy of the wavelengths of light emitted by hot bodies, he found that his formula fitted laboratory measurements exactly.

RELATED THEORIES
See also
THE ULTRAVIOLET
CATASTROPHE
page 16

EINSTEIN EXPLAINS THE
PHOTOELECTRIC EFFECT
page 20

BALMER'S PREDICTABLE
SPECTRUM
page 22

BOHR'S ATOM
page 24

3-SECOND BIOGRAPHY
NIELS BOHR
1885–1962
Danish pioneer of quantum theory and frequent opponent of Einstein

30-SECOND TEXT
Alexander Hellemans

Rather than continuous waves, Planck envisaged light as self-contained packets: quanta.

EINSTEIN EXPLAINS THE PHOTOELECTRIC EFFECT

the 30-second theory

3-SECOND FLASH
By suggesting that light is composed of energy packets called photons, Einstein was able to explain the puzzling features of the photoelectric effect.

3-MINUTE THOUGHT
Millikan spent ten years testing Einstein's theory in painstaking experiments that required extremely clean metal electrodes – but he did so because he was convinced the theory was wrong. Even when Millikan's results supported the predictions, he didn't believe Einstein's quantum view of light, saying that they lacked any 'satisfactory theoretical foundation'. Revolutionary ideas often do.

In 1905 it must have seemed as though Einstein was an unstoppable source of revolutionary ideas. Five years previously Planck had proposed that bodies emit electromagnetic radiation such as light in packets or 'quanta' with an energy proportional to their frequency. For Planck this hypothesis was a mathematical trick that made the equations produce sensible results. But when Einstein postulated that the quantization of energy was not some quirk of light emission but a fundamental property of light itself – that light was composed of a stream of discrete particles called photons, and not a continuous beam – it was a step too far for most scientists. However, Einstein suggested a way of testing his hypothesis. In the early 1900s Philipp Lenard had shown that light shone onto pieces of metal will eject electrons: the so-called photoelectric effect. But there was something odd: if the light was made brighter the electrons didn't escape with more energy, simply there were more of them. In Einstein's new model that made sense: a brighter beam contains more photons, albeit with the same energy as before. After Einstein's predictions about the photoelectric effect were verified experimentally by Robert Millikan over the ensuing decade, Einstein was awarded the 1921 Nobel prize in physics for his work.

RELATED THEORIES
See also
THE ULTRAVIOLET CATASTROPHE
page 16

PLANCK'S QUANTA
page 18

3-SECOND BIOGRAPHIES
PHILIPP LENARD
1862–1947
German experimental physicist, Nobel laureate in 1905, and Nazi sympathizer who referred to Einstein's work as 'Jewish physics'

ALBERT EINSTEIN
1879–1955
German-born physicist who developed special and general relativity and contributed to the origins of quantum theory

30-SECOND TEXT
Philip Ball

Einstein realized it was the energy of individual quanta of light that ejected electrons in photoelectric experiments.

BALMER'S PREDICTABLE SPECTRUM

the 30-second theory

When Niels Bohr was working on his quantum model of the atom, his aim was to provide a stable structure for electrons to exist around a central, positively charged nucleus. But in February 1913 he picked up on a discovery published by a schoolteacher, Johann Balmer, 28 years earlier. Bohr was chatting with a colleague, Hans Hansen, who mentioned that Balmer had produced a formula predicting the spectral lines emitted by hydrogen. When an element is heated it does not produce continuous colours, but narrow lines from the colour spectrum. Balmer had discovered that the frequency of these lines corresponded to a simple numerical formula. Until hearing this, Bohr had assumed that atoms emit light with frequencies corresponding to the rate of vibration or rotation of an electron – the accepted theory at the time. Balmer's equation caused Bohr to realize that the frequency of the light, linked to the energy of photons by Planck's simple formula, corresponded to the different energy gaps between the fixed electron orbits that he had devised. Bohr's new model not only explained why the atom was stable, but why specific frequencies of light were emitted in its spectrum.

3-SECOND FLASH
The accidental discovery of an old formula led Bohr to realize that his new atomic model explained both the stability of atoms and the energy of the photons they emitted.

3-MINUTE THOUGHT
Bohr told fellow physicist Léon Rosenfeld 'As soon as I saw Balmer's formula, the whole thing was immediately clear to me … I didn't know anything about the spectral formulae. Then I looked it up … And I found that there was this very simple thing about the hydrogen spectrum.' The formula was in a textbook Bohr had used as a student, so he would have seen it, but it was Hansen's casual remark that triggered Bohr's major contribution to atomic theory.

RELATED THEORIES
See also
THE ULTRAVIOLET CATASTROPHE
page 16

BOHR'S ATOM
page 24

THE DIRAC EQUATION
page 60

3-SECOND BIOGRAPHIES
JOHANN JAKOB BALMER
1825–98
Swiss schoolteacher who also lectured in mathematics at the University of Basel

LÉON ROSENFELD
1904–74
Belgian quantum physicist and documenter of the history of modern physics who named the lepton class of particle

30-SECOND TEXT
Brian Clegg

If electrons had fixed orbits, jumping between orbits would produce characteristic colours of light.

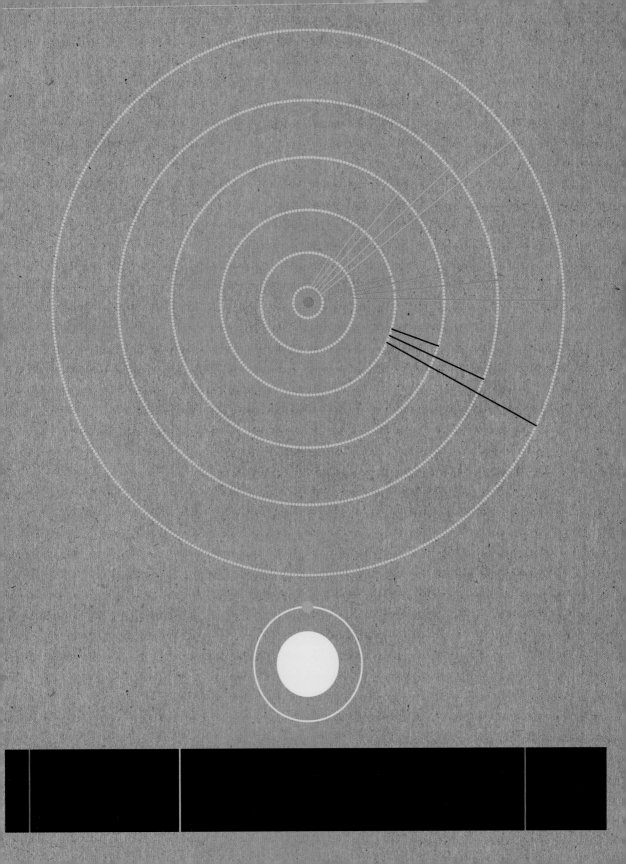

BOHR'S ATOM

the 30-second theory

When Niels Bohr came to

England in 1911 with a grant for a year's study, the structure of the atom was unknown. After moving to Manchester, where Ernest Rutherford had his lab, Bohr set out to provide a model that would work with Rutherford's discovery of the atomic nucleus. He needed to find a stable structure for the electrons in an atom to exist outside the nucleus. Bohr first suggested that electrons had an elastic connection to the nucleus, using Planck's idea of quanta to limit their possible frequencies of vibration, but this didn't match observations. It was already known that there was no stable structure with the electrons fixed in place, but the alternative, electrons that moved in orbits like planets, produced its own problems. When a charged body accelerates (as anything in orbit does), it gives off electromagnetic radiation. If electrons were planets around the sun of the nucleus they ought to spiral into the centre, losing energy as they went, and be destroyed. Because such an outcome clearly was not evident, Bohr suggested the electrons could only move in fixed orbits, jumping from orbit to orbit, never occupying the intervening spaces.

RELATED THEORIES
See also
PLANCK'S QUANTA
page 18

BALMER'S PREDICTABLE
SPECTRUM
page 22

EPR
page 98

3-SECOND BIOGRAPHIES
JOSEPH JOHN THOMSON
1856–1940
British discoverer of the electron

ERNEST RUTHERFORD
1871–1937
New Zealand-born discoverer of radioactive half-life and the atomic nucleus

30-SECOND TEXT
Brian Clegg

3-SECOND FLASH
In his model of the atom, Bohr put the electrons on orbital tracks, preventing them from spiralling to their destruction.

3-MINUTE THOUGHT
A strong influence on Bohr's thinking was what he described as 'the radiation mechanism proposed by Planck and Einstein' – the idea that atoms could only emit light in fixed chunks or 'quanta'. By putting electrons in set orbits, he could ensure that the energy they absorbed by jumping up a level, or released when dropping down a level – known as a quantum leap – corresponded to the energy of an appropriate photon.

Unlike planets, electrons in Bohr's atom were limited to specific orbits, unable to occupy the spaces in between.

7 October 1885
Born in Copenhagen, son
of Danish Professor of
Physiology Christian Bohr
and his Jewish wife Ellen,
née Adler

1908
Publishes prize-winning
paper on surface tension
in *Transactions of the
Royal Society*

1911
Awarded his doctorate at
Copenhagen University

1911–12
Year in England at
Cambridge and Manchester
sets him on the path to
the quantum atom

1912
Marries Margrethe
Norlund. She becomes
his secretary

1913
Publishes 'Bohr atom'
model

1913
Lectureship in Physics at
Copenhagen University

1914
Lectureship in Physics at
Manchester University

1916
Becomes Professor of
Theoretical Physics at
Copenhagen University

1920
Appointed Head of the
new Institute of
Theoretical Physics,
Copenhagen University

1922
Receives Nobel Prize in
Physics for his atomic
structure

mid 1920s
Development of modern
quantum mechanics, the
so-called Copenhagen
Interpretation, became
the basis for a series of
intense debates between
Bohr and Einstein

1931
Moves with his family
into the Carlsberg
Honorary Residence in
Valby, Copenhagen

early 1930s
Becomes interested in
the atom's nucleus

1943
Escapes arrest by German
police by crossing to
Sweden, subsequently
the UK and then the
United States. Consulted
on the atomic bomb

18 November 1962
Dies in Copenhagen

1965
Danish Institute of
Theoretical Physics is
renamed Niels Bohr
Institute

1997
Element 107 named
bohrium

NIELS BOHR

Impressive as Niels Bohr's freshly minted doctorate from Copenhagen was, the young Danish physicist transformed his career when, in 1911, he set off for a year in the UK, beginning in Cambridge under J. J. Thomson, discoverer of the electron. A first meeting with Thomson, during which Bohr told him there was a mistake in one of his books, may have soured relations, but Bohr was soon invited to join Ernest Rutherford in Manchester, where he built on Rutherford's discovery of the atomic nucleus to develop a model of the structure of the atom based on the new quantum theory. From the publication of Bohr's papers on the atom, he never looked back.

Brilliant as he was, Bohr could appear slow-thinking. He would spend a vast amount of time structuring his ideas before speaking. A colleague, James Franck, said: 'His face became empty, his limbs were hanging down, and you would not know this man could even see. You would think that he must be an idiot. There was absolutely no degree of life. Then suddenly one would see that a glow went up in him and a spark came, and then he said: "Now I know."'

Bohr was at the heart of the development of quantum theory, alongside Schrödinger, de Broglie and Heisenberg. He dismissed Einstein's concerns about the nature of this powerful but mysterious theory explaining the behaviour of atoms, electrons and photons.

Einstein hated the idea that probability was central to quantum theory, and sprang traps on Bohr when they met at conferences, preparing complex thought experiments that he believed demonstrated flaws in the theory. Bohr would usually spend a day thinking through the problem and come back with a solution.

For many years Bohr headed up the Institute of Theoretical Physics in Copenhagen, where he worked on the nature of quantum theory, devising the idea of 'complementarity', which suggests that the way quantum particles are measured will inevitably influence the results. In 1931 he was invited to live in the Carlsberg Honorary Residence in Copenhagen. The villa, home of the founder of Carlsberg, brewer Carl Jacobsen, was made available to the person who brought the greatest honour to Denmark.

In the mid-1930s, Bohr extended Carl von Weizsäcker's 'liquid drop' model of the nucleus, which treats the protons and the then newly discovered neutrons as an incompressible fluid, providing predictions on the binding energy of the nucleus that would be valuable to the team who discovered nuclear fission. By the mid-1940s, Bohr, whose mother was Jewish, was at risk in Nazi-occupied Denmark and he was forced to make his home in the UK and then the United States. Shy and sometimes diffucult to understand, Bohr inspired a generation of students to take further strides in physics.

Brian Clegg

WAVE–PARTICLE DUALITY

the 30-second theory

3-SECOND FLASH
De Broglie, by assuming that particles can behave as waves and waves as particles, resolved a paradox that was plaguing quantum theory.

3-MINUTE THOUGHT
Experimenters soon discovered that it was possible either to measure the properties of light as particles or as waves, but not both. This led Bohr to formulate his complementarity principle: light appears as particles when its energy is measured, for example, in the photoelectric effect or as a wave in diffraction experiments. The complementarity principle, which also includes Heisenberg's uncertainty principle, independently was the basis of the Copenhagen interpretation of quantum mechanics.

In 1905 Einstein explained the photoelectric effect – the emission of electrons from metallic surfaces when irradiated with light – by assuming that light consists of particles that kick out electrons from metals. Einstein realized that these light particles were the quanta discovered by Planck. However, the idea of light particles, later called photons, was seen as contradictory to the behaviour of light as a wave in phenomena such as diffraction – white light is spread out into the colours of the rainbow, as seen in reflected light on a CD – or interference – the fact that two light beams created by a double slit can cancel each other out. This contradiction preoccupied Einstein until 1923 when Louis de Broglie postulated the idea that if light photons can behave either as particles or as waves, the same should also be true of other particles, such as electrons. Erwin Schrödinger subsequently picked up de Broglie's idea and by looking at electrons in atoms as waves he calculated the wavelengths of the emitted light. In 1927, George Paget Thomson and Clinton Davisson showed that a narrow electron beam, when it passes through a thin metal film or crystal, spreads out in a diffraction pattern of concentric rings, thus proving de Broglie's theory.

RELATED THEORIES
See also
PLANCK'S QUANTA
page 18

EINSTEIN EXPLAINS THE PHOTOELECTRIC EFFECT
page 20

BOHR'S ATOM
page 24

DE BROGLIE'S MATTER WAVES
page 30

3-SECOND BIOGRAPHIES
ERWIN SCHRÖDINGER
1887–1961
Austrian physicist and pioneer of wave mechanics

LOUIS DE BROGLIE
1892–1987
French physicist who realized that any particles can behave as waves

30-SECOND TEXT
Alexander Hellemans

Depending on the process and the way it is measured, quantum entities can appear to be waves or particles.

DE BROGLIE'S MATTER WAVES

the 30-second theory

If, as Einstein had shown, waves of light could behave like particles of matter, could particles of matter behave like waves? In 1923, Louis de Broglie set out to see whether this could be true and developed a theory of matter waves. He realized that if particles such as electrons were to behave like waves they needed to have a frequency and a wavelength – the key measurable properties of wave – and to produce diffraction and interference patterns – phenomena exclusively related to waves. De Broglie calculated that the wavelength of a particle of known mass is related to its speed and that the particle's frequency is proportional to its energy. His ideas were confirmed by the discovery of electron diffraction in separate experiments by George Thomson in the UK, and Clinton Davisson and Lester Germer in the United States in 1927. Both experiments showed that electrons fired at a solid target behaved like waves and exhibited diffraction. Subsequent demonstration of electron interference in the quantum double slit experiment provided further confirmation. Diffraction and interference patterns have since been shown for progressively larger molecules – some big enough to be viewed under electron microscopes. De Broglie's discovery of the wave nature of electrons broke new ground in the development of quantum physics.

3-SECOND FLASH
De Broglie showed that particles could behave like waves just as waves could behave like particles.

3-MINUTE THOUGHT
A particle's wavelength, known as its de Broglie wavelength, is inversely proportional to the momentum of the particle. It can be calculated for an atom, a molecule and, in principle, for a much bigger object, though because the mass of, say, a person is so large compared with a particle, its wavelength is too small to be observed.

RELATED THEORIES
See also
EINSTEIN EXPLAINS THE PHOTOELECTRIC EFFECT
page 20

WAVE-PARTICLE DUALITY
page 28

THE QUANTUM DOUBLE SLIT
page 32

3-SECOND BIOGRAPHIES
CLINTON DAVISSON
1881–1958
American physicist who discovered electron diffraction in a nickel crystal

GEORGE PAGET THOMSON
1892–1975
British physicist who discovered electron diffraction in a thin metal sheet

LOUIS DE BROGLIE
1892–1987
French physicist who proposed the wave nature of electrons

30-SECOND TEXT
Leon Clifford

Electron diffraction patterns show that electrons are behaving like waves rather than conventional particles.

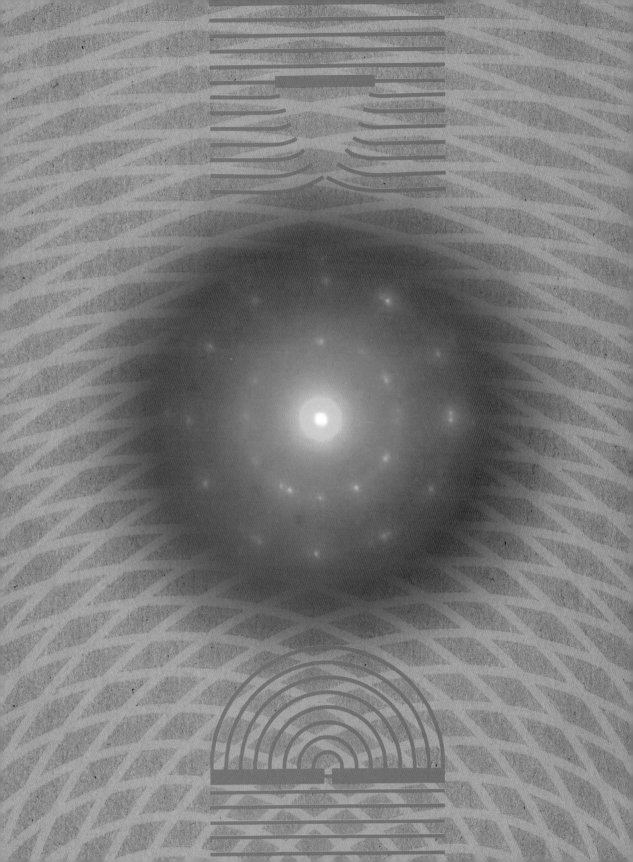

QUANTUM DOUBLE SLIT

the 30-second theory

3-SECOND FLASH
The double-slit experiment demonstrates the fundamentally dual nature of light, which is both a wave and a stream of particles.

3-MINUTE THOUGHT
If photon detectors are inserted into the light path to allow the experimenter to determine which of the two slits a particular photon goes through, then the interference pattern disappears. The light is now behaving like particles, not waves. In Wheeler's 'delayed choice' version of the experiment, the decision to look for particle-like or wave-like behaviour is made after the photon has passed through the slit … but you still see whichever type of behaviour you are looking for.

A century before Planck and Einstein demonstrated that light is emitted and absorbed as particle-like photons, Thomas Young devised a famous experiment that seemed to show something different. If a beam of light is passed through two narrow slits and projected onto a screen, you might expect to see just two bright lines opposite the slits. Instead, the experiment produces a pattern of multiple, closely spaced lines. These lines, called interference fringes, are characteristic of waves, not particles. Yet the light hitting the screen is detected, as light always is, in the form of discrete photons. The double-slit experiment is a dramatic demonstration of the dual nature of light: it propagates like a wave, but interacts like a stream of particles. The same is true of matter: the double-slit experiment yields identical results if electrons are used instead of photons. Even if the electrons are fired through the slits one at a time, the same characteristic interference pattern builds up. A single electron acts like a wave that can interfere with itself! Originally a 'thought experiment' attributed to Richard Feynman, who famously described the double slit as 'the heart of quantum mechanics', this bizarre result has now been verified in the laboratory.

RELATED THEORIES
See also
WAVE–PARTICLE DUALITY
page 28

COLLAPSING WAVE FUNCTIONS
page 50

BEAM SPLITTERS
page 78

QUANTUM OPTICS
page 132

3-SECOND BIOGRAPHIES
THOMAS YOUNG
1773–1829
English polymath and brilliant pioneer in diverse subjects

RICHARD FEYNMAN
1918–88
American theoretical physicist and Nobel prizewinner who was also a great science communicator

30-SECOND TEXT
Andrew May

Classically waves are required to produce interference from the double-slit experiment.

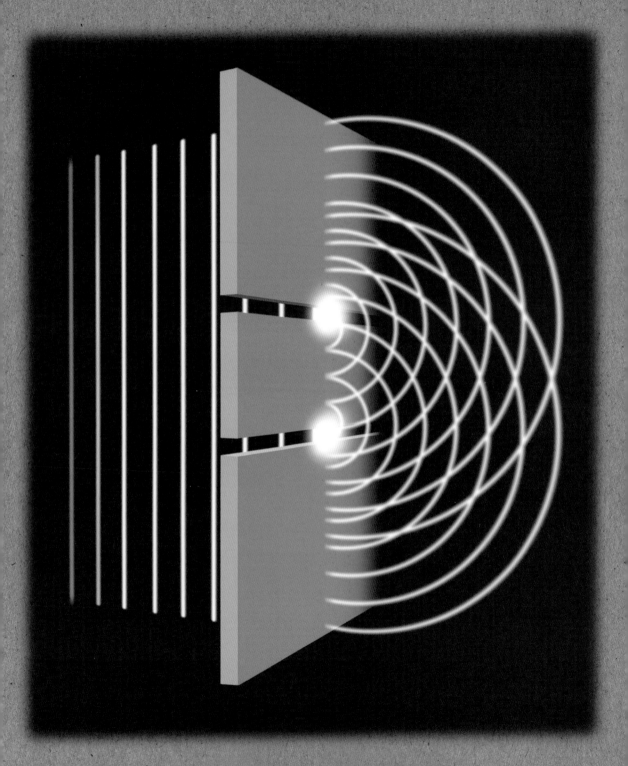

THE ESSENTIALS

angular momentum Momentum is the 'oomph' of a travelling body – mass times velocity. Angular momentum is the equivalent for a rotating body, combining momentum with the body's distance from the centre of rotation.

classical physics The physical theories that held sway before 1900 and the twin twentieth century upheavals of relativity and quantum theory. Newton's laws of motion is a typical example of a classical theory, which would be superseded by special relativity, but still is a good approximation for most moving bodies.

complementary variables Heisenberg's Uncertainty Principle links pairs of a quantum particle's properties. The best known of these complementary paired variables are position with momentum, and energy with time. The more accurately one variable is known, the less accurately the other.

matrix A collection of numbers in a regular array. Often rectangular, matrices can have any number of dimensions. Matrices are used to work on multiple equations simultaneously.

non-relativistic equation An equation that does not take relativity into account. Newton's second law (force = mass x acceleration) is a non-relativistic equation. For speeds well below the speed of light this is effectively correct, but as velocity increases, relativistic effects become important as, for instance, the mass of an object increases with velocity.

particle accelerators The main tool of particle physics. Accelerators push charged particles close to the speed of light then slam them into other particles or solid objects. The result is a spray of new particles generated by the collision. To date, the biggest such machine is the Large Hadron Collider (LHC) at CERN. Spanning the Swiss/French border, the LHC is in a 17-mile (27-kilometre) tunnel that accelerates streams of protons in opposite directions before colliding them.

quantum states The set of values of a quantum particle's properties. A state can be 'pure', where it has a specific value – for example, the spin when measured will be 'up' or 'down' – or mixed, in which case the spin might be 40 per cent probability of 'up' and 60 per cent probability of 'down'.

quantum electrodynamics Usually shortened to QED, quantum electrodynamics is our theory of how light and matter (usually an electron) interact. It is a relativistic quantum field theory, because it takes into account special relativity, is quantized and represents a field by imagining each particle accompanied by a rapidly spinning clock. Its hand indicates the particle's phase and the probability of taking a particular path.

spacetime Relativity treats time as a fourth dimension. In relativity there is no absolute position or absolute time because the way things move influences their position in time, so it is necessary to consider spacetime as a whole, rather than to think of space and time independently.

superposition When a quantum particle has a state with, say, two possible values it will not have an actual value but rather a superposition – a collection of probabilities of being in the states, until it is measured, when it collapses to an actual value. A tossed coin has two states but no superposition. Before we look, the coin already has one of these values. But a quantum particle has no value, literally just probabilities, while in superposition.

virtual particles QED requires virtual particles, which are never seen but take part in quantum processes. The electromagnetic force, for instance, causes an electron to change path because the electron absorbs a virtual photon. Also in empty space, the uncertainty principle means that it is possible for energy levels to fluctuate, briefly bringing a pair of virtual particles – matter and antimatter – into existence before disappearing again to energy.

wavefunction In quantum physics, the wavefunction is a mathematical formula that describes the behaviour of a quantum state of the particle, which evolves over time according to the Schrödinger wave equation. The wave in question, which spreads out over time, does not describe the particle itself, but rather the probability of a quantum state having a particular value – so, for instance, it can describe the probability of finding a particle in different locations. The probability is given by the square of the wavefunction.

QUANTUM SPIN

the 30-second theory

Quantum spin is the cause of magnetism in the everyday world. It is an intrinsic property of subatomic particles, a key parameter needed to describe a particle fully and one of four attributes required to define the quantum state of an electron in an atom. Quantum spin, just like everything else in quantum mechanics, comes in fixed packets and particular particles can have only certain amounts of spin, which is represented by their quantum spin number. All sub-atomic particles have a quantum spin number although some can have a spin number of zero. Quantum spin is related to angular momentum, the physical property attributed to rotating objects, in that it affects the measurement of angular momentum in atoms. The effects of quantum spin were first detected in relation to electrons in atoms. Electrons whizzing around the nucleus of atoms impart angular momentum to those atoms through their orbital motion. Quantum spin was discovered by German physicists Otto Stern and Walther Gerlach in 1922 during an experiment that suggested electrons in atoms also have some intrinsic angular momentum in addition to that generated by their orbital motion. This is akin to an electron spinning on its own axis while it is orbiting the atomic nucleus.

RELATED THEORIES
See also
THE PAULI EXCLUSION
PRINCIPLE
page 58

THE DIRAC EQUATION
page 60

QUANTUM FIELD THEORY
page 64

3-SECOND BIOGRAPHIES
WOLFGANG PAULI
1900–58
Pioneering Austrian physicist who developed much of the theory on quantum spin

GEORGE UHLENBECK &
SAMUEL GOUDSMIT
1900–88 & 1902–78
Dutch physicists who co-wrote the first paper on electron spin

30-SECOND TEXT
Leon Clifford

3-SECOND FLASH
Quantum spin is the reason why magnets work and it allows scientists to distinguish particles from one another.

3-MINUTE THOUGHT
The property we call (quantum) spin was given the name because it has some similarities to classical angular momentum, but we have no reason to think that this is to do with particles spinning around (which is hard to envisage for a point particle like an electron), especially as spin's half integer values are only ever found to be 'up' or 'down' in any direction in which they are measured.

The direction of quantum spin of particles determines their magnetic orientation.

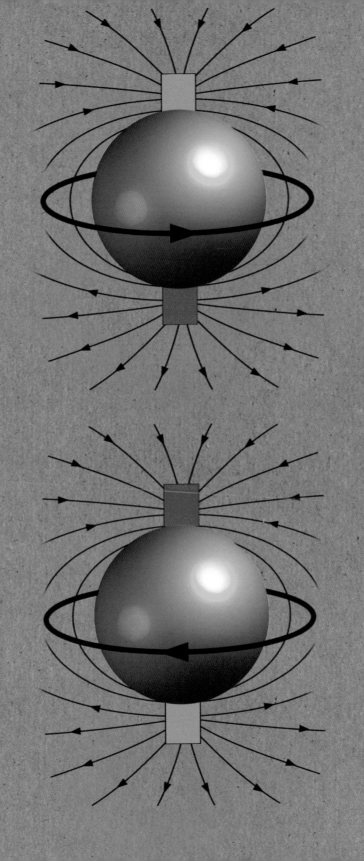

MATRIX MECHANICS

the 30-second theory

3-SECOND FLASH
The principle behind matrix mechanics helped Heisenberg and his colleagues to put the uncertainty into quantum mechanics.

3-MINUTE THOUGHT
Heisenberg's matrices offered physicists a new and unfamiliar way of describing quantum behaviour that had no equivalent in the world we observe. But it seemed to be at odds with the more traditional approach, preferred by Schrödinger among others, of using differential equations. In one sense, this was another manifestation of wave-particle duality; differential equations have smoother wave-like features while matrices appear more like discrete objects. Dirac combined the two into a single formalism in 1930.

Matrix mechanics is a way of describing the behaviour of quantum systems using a particular mathematical technique known as matrix algebra. The approach was developed in 1925 by German physicists Werner Heisenberg, Max Born and Pascual Jordan who were trying to understand the rules governing atomic spectra. It contrasts with the approach taken by Erwin Schrödinger who used a different mathematical technique involving differential equations to describe quantum systems. The significance of matrices lies in the fact that the order in which things are done affects the outcome. In everyday mathematics, for example, the multiplication of two numbers gives the same result no matter which way round the numbers are ordered: 2 x 3 is the same as 3 x 2. This is not the case with matrices. Instead, if the position of a particle is represented by one matrix and if the momentum of that same particle is represented by another matrix, the multiplication of these two matrices will give different results depending on the order in which it is done. The result of multiplying the position matrix by the momentum matrix is not the same as multiplying the momentum matrix by the position matrix. The difference between the two results gives rise to Heisenberg's Uncertainty Principle.

RELATED THEORIES
See also
SCHRÖDINGER'S EQUATION
page 42

HEISENBERG'S UNCERTAINTY PRINCIPLE
page 48

COLLAPSING WAVEFUNCTIONS
page 50

3-SECOND BIOGRAPHIES
MAX BORN & PASCUAL JORDAN
1882–1970 & 1902–80
German physicists who developed the mathematics of matrix mechanics

WERNER HEISENBERG
1901–76
German quantum pioneer who used maths to reveal the uncertain and bizarre nature of the quantum world

30-SECOND TEXT
Leon Clifford

Heisenberg's matrix mechanics used the mathematics of matrices to predict quantum behaviour.

$$q = \begin{pmatrix} q_{11} & q_{12} & q_{13} & \cdot\cdot \\ q_{21} & q_{22} & q_{23} & \cdot\cdot \\ q_{31} & q_{32} & q_{33} & \cdot\cdot \\ \cdot\cdot & \cdot\cdot & \cdot\cdot & \cdot\cdot \end{pmatrix}; \quad p = \begin{pmatrix} p_{11} & p_{12} & p_{13} & \cdot\cdot \\ p_{21} & p_{22} & p_{23} & \cdot\cdot \\ p_{31} & p_{32} & p_{33} & \cdot\cdot \\ \cdot\cdot & \cdot\cdot & \cdot\cdot & \cdot\cdot \end{pmatrix}$$

SCHRÖDINGER'S EQUATION

the 30-second theory

3-SECOND FLASH
Schrödinger's equation
provides a way to calculate
how quantum particles
behave as 'probability
waves': describing where
they are likely to be found
at any moment.

3-MINUTE THOUGHT
Schrödinger's 'wave
mechanics' isn't the
only way to write
down quantum theory
mathematically. While
Schrödinger was conjuring
up his equation in the
1920s, Werner Heisenberg
was also developing a
way to express quantum
states as tables of
numbers called matrices.
This 'matrix mechanics'
is still sometimes used,
but Schrödinger's waves
are generally preferred,
partly because they offer
a more intuitive picture
of the 'appearance' of
quantum states.

Louis de Broglie's proposal in 1924 that particles such as electrons could behave as waves stimulated Erwin Schrödinger to formulate a mathematical theory of quantum mechanics in which the position and behaviour of particles are described in terms of a wavefunction ψ ('psi'). This is a kind of wave, but not in the usual sense of a sound wave or an ocean wave. Rather, the wavefunction is a probability wave: its value (or, more properly, the value of the square of the wave function $\psi 2$) at any point in space indicates the probability of finding the particle there. Waves are described mathematically by so-called differential equations, which specify how the size of the oscillations changes over time. But Schrödinger's equation is unlike an ordinary wave equation, being more akin to the kind of equation used to describe spreading or diffusion processes. In principle, it enables scientists to calculate the wavefunction of any quantum system, and thus the probabilities of its locations, provided its mass and energy are known. In practice, solving the equation exactly is often too hard, and only approximate solutions can be found. Nonetheless, Schrödinger's equation is the starting point for all attempts to work out how electrons are distributed in atoms, molecules and materials.

RELATED THEORIES
See also
WAVE–PARTICLE DUALITY
page 28

DE BROGLIE'S MATTER WAVES
page 30

MATRIX MECHANICS
page 40

3-SECOND BIOGRAPHY
ERWIN SCHRÖDINGER
1887–1961
Austrian physicist whose work
embraced quantum theory,
cosmology and genetics

30-SECOND TEXT
Philip Ball

Probabilities for locating hydrogen's electron in different orbits can be predicted by its wavefunction.

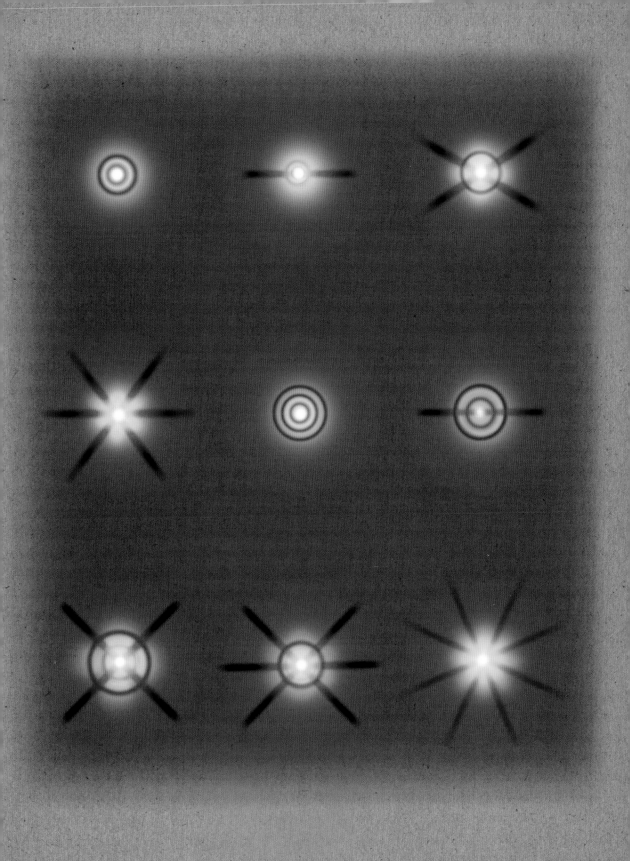

12 August 1887
Born in Vienna, Austria

1910
Awarded a doctorate by the University of Vienna

1914–18
Serves as an artillery officer in the Austrian Army during the First World War

1921
Appointed Professor of Theoretical Physics at Zurich, Switzerland

1926
Schrödinger's equation lays the foundations of wave mechanics

1927
Moves to Berlin, taking up the professorship vacated by Max Planck

1933
Voluntarily leaves Nazi Germany and moves to Oxford; in the same year shares the Nobel Prize in Physics with Englishman Paul Dirac

1935
A paper entitled 'The Present Situation in Quantum Mechanics' presents the paradox of Schrödinger's Cat

1939
Appointed Director of Theoretical Physics at the Institute for Advanced Studies, Dublin, Ireland

1944
Cambridge University Press publishes *What is Life?*

1956
Retires his post in Dublin and returns to Vienna

4 January 1961
Dies in Vienna

ERWIN SCHRÖDINGER

Born and raised in Vienna, Erwin Schrödinger was a brilliant young student with an instinctual flair for physics. By the time he secured his first permanent professorship, in Zurich in the 1920s, he had become a firm believer in the wave nature of matter. This culminated in the theory of wave mechanics – a complete and self-consistent formulation of quantum theory that is generally considered to be his greatest achievement.

Schrödinger's wave equation is as central to quantum physics as are Newton's laws of motion to classical physics. The mathematical validity and predictive power of Schrödinger's equation were immediately recognized by his fellow physicists, but to his dismay virtually none of them shared his unmitigated enthusiasm for waves. To him, the prevailing Copenhagen Interpretation, championed by Niels Bohr, with its talk of 'wave-particle duality' and 'collapse of the wavefunction' was pseudoscientific nonsense. Encouraged by Einstein, one of the few people at the time to share his view, Schrödinger set about devising a thought-experiment that would highlight the absurdity of the Copenhagen Interpretation. The result was the paradox of Schrödinger's Cat – probably the best-known image in all of quantum theory.

Before the outbreak of the Second World War, Schrödinger escaped the turmoil of continental Europe for neutral Ireland. His maternal grandmother had been British, and he spoke English almost as fluently as he spoke German. At the personal invitation of Éamon de Valera, then Irish Prime Minister, Schrödinger became Director of Theoretical Physics at the newly formed Institute for Advanced Studies in Dublin – a post he held for the next 17 years. He later described his time in Dublin as the happiest years of his life. Arguably Schrödinger's most important work during this period was *What Is Life?* – a revolutionary little book that demonstrated how quantum theory and other concepts from fundamental physics could be applied to living organisms. When the secrets of DNA were unlocked a few years later by Francis Crick and James Watson, both men acknowledged their debt to Schrödinger's book.

Compared with his scientific peers, Schrödinger had an unusual lifestyle. He wrote poetry and had an abiding interest in philosophy and Eastern mysticism. More controversially, he had a string of young mistresses. His three acknowledged children were all born during his 40-year marriage to his wife Anni, but none were hers. Anni seems to have been resigned to her husband's serial infidelity, however, and she remained his wife until his death in Vienna in 1961 at the age of 73.

Andrew May

SCHRÖDINGER'S CAT

the 30-second theory

The most famous illustration of how quantum theory defies intuition is a thought experiment proposed in 1935 by Erwin Schrödinger. He suggested that it should be possible to make the state of a macroscopic object – whether a cat enclosed in a box is alive or dead, say – dependent on a microscopic quantum event, such as the decay of an atom. He imagined a device in which the radioactive decay of an atom – a quantum event governed by chance – releases a hammer that breaks a vial of poison, killing the cat. The problem is that the decaying atom can be in a mixture of states, called a superposition, implying that the cat may be simultaneously both killed and not killed. Quantum superpositions are generally destroyed by making a measurement on the quantum object, so immediately we open the box to look, the cat will be in one state or the other. But that doesn't tell us the cat's condition before we looked. Some scientists feel that something will intervene to put the cat in one state or the other, whether we look or not. Others are content to imagine a live-dead superposition for the cat.

RELATED THEORIES
See also
SCHRÖDINGER'S EQUATION
page 42

COLLAPSING
WAVEFUNCTIONS
page 50

DECOHERENCE
page 52

3-SECOND BIOGRAPHIES
EUGENE WIGNER
1902–95
Hungarian-born physicist who linked Schrödinger's cat to the problem of consciousness by postulating a friend who performed the experiment in Wigner's absence

JUAN IGNACIO CIRAC
1965–
Spanish physicist who proposed a Schrödinger-cat experiment using microscopic living organisms

30-SECOND TEXT
Philip Ball

3-SECOND FLASH
The Schrödinger's Cat thought experiment shows how counterintuitive quantum theory is: a quantum system that is simultaneously in two states determines whether a cat is alive or dead.

3-MINUTE THOUGHT
Could we test experimentally whether Schrödinger's cat is alive or dead? Sustaining a delicate quantum superposition of states in a system big enough to contain a real cat would be almost impossible, but a microscopic 'cat' – a bacterium or virus, say – could be isolated from disturbances more easily. Researchers in Germany have proposed an experiment in which a virus trapped by laser light could be coaxed into a quantum superposition.

Before observation, Schrödinger's cat is in a superposition of alive and dead.

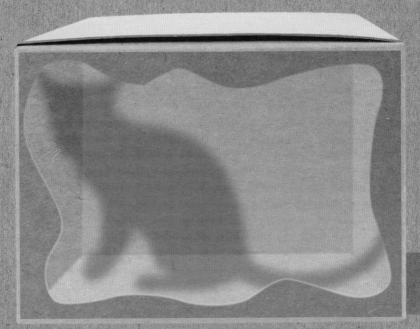

HEISENBERG'S UNCERTAINTY PRINCIPLE

the 30-second theory

3-SECOND FLASH
Particles are like politicians: the more you attempt to pin them down, the faster they change their position.

3-MINUTE THOUGHT
The Uncertainty Principle is a reason why particle accelerators, such as the Large Hadron Collider, are so huge. In order to handle distances a thousand times smaller than the size of a proton, we require beams of particles whose energies are trillions of times larger than those of particles at room temperature. This requires huge accelerators to energize the beams to such extremes.

In 1927 German theoretical physicist Werner Heisenberg formulated the Uncertainty Principle – a fundamental property of quantum systems. It states that it is impossible to measure simultaneously with perfect accuracy certain pairs of physical properties (so-called complementary variables) of an atom or a particle – for example, both its position and momentum – or to be certain of its energy at some specific instant in time. The more precise the measurement of one quantity, the less precisely can the other be measured or controlled. The effect of this phenomenon is so small that it can be ignored in everyday affairs, but it is dramatic for subatomic particles and underpins quantum mechanics, which describes the motion and dynamics of atoms. This uncertainty is an intrinsic limit on our ability to measure natural phenomena at small distances; it is a fundamental property of quantum theory and not simply a failure in the measuring apparatus. One consequence is that the total energy of a particle can fluctuate by some amount, E, for a short time, t, as long as the product of E times t does not exceed Planck's constant divided by 4 pi. This in turn means that the law of conservation of energy can be evaded for very short time spans.

RELATED THEORIES
See also
SCHRÖDINGER'S EQUATION
page 42

THE DIRAC EQUATION
page 60

FEYNMAN DIAGRAMS
page 70

3-SECOND BIOGRAPHIES
ERWIN SCHRÖDINGER
1887-1961
Austrian physicist who created a non-relativistic equation for quantum mechanics and the equation including relativity

WERNER HEISENBERG
1901-76
German theoretical physicist who proposed the Uncertainty Principle

30-SECOND TEXT
Frank Close

The closer we pin down the location of a quantum particle, the less we can know about its momentum (and vice versa).

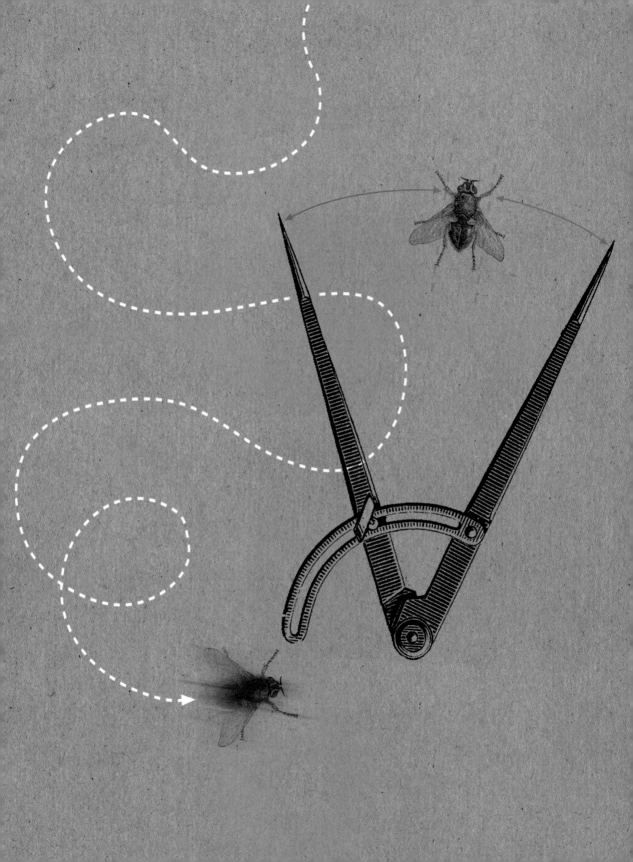

COLLAPSING WAVEFUNCTIONS
the 30-second theory

3-MINUTE THOUGHT
Another interpretation of wavefunction collapse was developed by American physicist David Bohm, based on de Broglie's 'pilot-wave' theory in which quantum particles are accompanied and guided by waves. It assumes that there is a single wavefunction governing the entire universe (so that everything depends on everything else), which never in fact collapses but just appears to do so locally, due to decoherence acting between the local wavefunction and that of the rest of the universe.

The Schrödinger equation, which encapsulates all we can know about a given quantum system in the form of the wavefunction, can only predict the various probabilities of finding it in a particular state, whereas a measurement made on the system gives a unique answer: we find it in *this* state or *that*. It appears that the very act of observing the system winnows the possibilities to a single outcome. This is called collapsing the wavefunction. Exactly how collapse occurs may depend on how the measurement is made. This represented one of the key philosophical shocks of the theory when it was first developed, because it seemed to undermine science's supposed objectivity: the observer, apparently, could not help but influence the result. This is now known as the 'measurement problem'. But is this collapse mere mathematical formality, or a real physical process? The conventional 'Copenhagen Interpretation' of quantum theory only insists that the state cannot be known until it is observed. Some think wavefunction collapse is an illusion, because all possible outcomes are realized in different worlds. For others collapse is a process like radioactive decay, with a definite timescale, which might involve the gravitational force, thus achieving the long-sought link between gravity and quantum theory.

RELATED THEORIES
See also
SCHRÖDINGER EQUATION
page 42

SCHRÖDINGER'S CAT
page 46

DECOHERENCE
page 52

CONSCIOUSNESS COLLAPSE
page 90

3-SECOND BIOGRAPHY
ROGER PENROSE
1931–
British mathematical physicist who proposes that wavefunction collapse is the result of spacetime curvature in general relativity

30-SECOND TEXT
Philip Ball

In a quantum system, like the two slit process, particles exist as a probability wave, causing interference and other effects: on observation, this collapses leaving a single value.

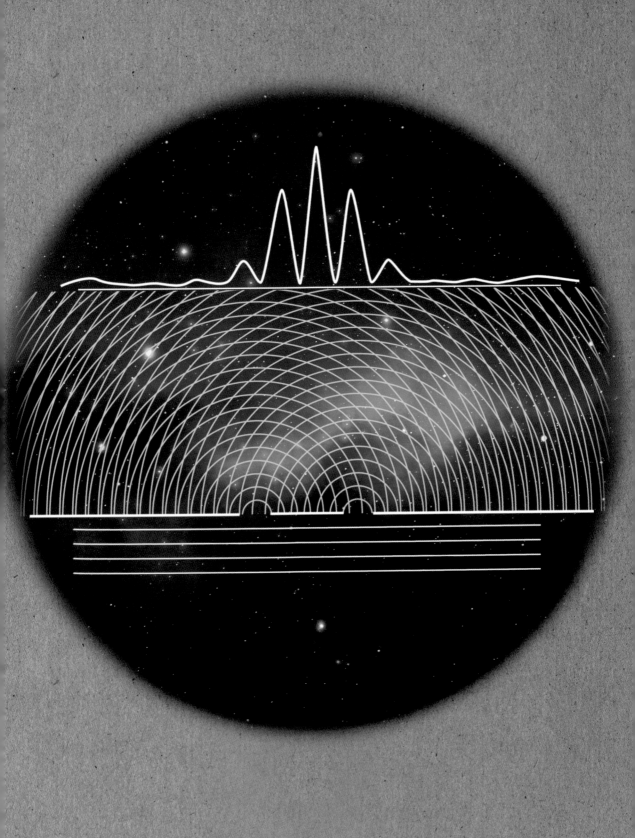

DECOHERENCE

the 30-second theory

3-SECOND FLASH
Decoherence is the loss of quantum behaviour owing to interactions of its constituent particles with their environment.

3-MINUTE THOUGHT
How big can a quantum system be before decoherence becomes impossible to suppress, and the system starts to behave classically? Some large molecules, such as 60-atom C_{60}, can show wavelike quantum interference effects, but these can be washed away by passing the molecules through a gas so that collisions induce decoherence. Soon it should be possible to detect quantum superpositions of vibrating states in tiny oscillating beams visible in the electron microscope.

The microscopic world is governed by quantum rules, but in the everyday world of billiard balls and teapots, classical physics applies. How does quantum become classical – where does the quantum weirdness go? A common view is that quantum effects such as the wave behaviour of particles get 'washed out' by interactions between the quantum particles and their environment, a phenomenon known as decoherence. These interactions mean that a particle and its environment become 'entangled': the properties of the particle are no longer intrinsic to it but depend on the environment. To see quantum behaviour in a system, decoherence must be suppressed, by making the system as isolated as possible from its environment. That is why quantum effects such as superpositions are usually observed only in the laboratory; they are fragile and easily destroyed by decoherence. Decoherence is a one-way affair: once it has washed away quantum-ness, you cannot get it back. The rate of decoherence – the speed at which quantum superpositions vanish – increases exponentially as the number of particles in the system increases, so big objects become classical almost instantly. Decoherence, then, makes the quantum-to-classical switch a well-defined process that depends on the precise environmental conditions.

RELATED THEORIES
See also
SCHRÖDINGER'S CAT
page 46

COLLAPSING
WAVEFUNCTIONS
page 50

3-SECOND BIOGRAPHIES
HEINZ-DIETER ZEH
1932–
German physicist who in 1970 identified the origin of decoherence

WOJCIECH ZUREK
1951–
Polish-American physicist who explained how decoherence 'selects' a few classical properties from the palette of possible quantum states

30-SECOND TEXT
Philip Ball

Quantum particles have superposed states, but decoherence ensures that everyday items stay firmly classical.

THE PHYSICS OF LIGHT & MATTER

THE PHYSICS OF LIGHT & MATTER
GLOSSARY

antimatter British physicist Paul Dirac predicted that the electron should have an equivalent with a positive charge, later called a positron. This was discovered to exist; the first example of an antimatter particle; subsequently further examples were found for all matter particles. When energy converts into matter it produces a pair of matter and antimatter equivalents, which can recombine, annihilating to energy.

bosons A particle that obeys Bose-Einstein statistics (as opposed to a fermion). Typically bosons are the particles that carry forces, most notably photons and the famous Higgs boson, but the term also applies to atomic nuclei with an even numbers of particles. Unlike a fermion, many bosons can be in the same state at the same time.

divergent series A series in which the sum is infinite. $1 + \frac{1}{2} + \frac{1}{3} + \frac{1}{4} + \frac{1}{5}$ ... (where '...' means 'continue this series forever') is divergent. By contrast, a convergent series has a finite sum. The total of $1 + \frac{1}{2} + \frac{1}{4} + \frac{1}{8}$ ... is just 2, even though it contains an infinite set of fractions, because each subsequent item takes the total closer to 2, but never exceeds it.

electron shells Electrons only occupy fixed orbits around an atom, with jumps between these orbits usually involving absorbing or giving off a photon – this is the quantization of the electron. The possible levels of orbit are sometimes called shells, most often in chemistry. Each shell holds a maximum number of electrons (2 in the first shell, 8 in the second, 18 in the third and so on).

fermions One of the two principal types of particle (the other being bosons). Matter particles (electrons, quarks, protons and neutrons) and neutrinos are fermions. Atoms with an odd number of fermions are also fermions, while those with an even number are bosons. Fermions obey Fermi–Dirac statistics, and the Pauli Exclusion Principle (see page 58), which prevents more than one being in an identical state.

fields (quantum fields) A field is a mathematical construct filling all spacetime with a value at every location – a bit like a three-dimensional map of the Earth, where the height above sea-level is the field value. A quantum field is one producing the same effects as quantum objects that can be in a superposition of states, requiring more complex mathematics than a classical field like a map of the Earth.

fine-structure constant One of the fundamental constants of physics, with a value of about $\frac{1}{137}$. The fine-structure constant (a) reflects the strength of the electromagnetic attraction (in effect, the probability an electron will emit a photon) and controls the way electrons bind in atoms and molecules.

matrix mechanics An early formulation of quantum theory by Heisenberg that made no attempt to provide an illustrative picture, but merely predicted what was observed as the outcome of the changing values of a set of numbers over time.

neutrons Neutrally charged particles found in the nucleus of atoms, consisting of three quarks. A particular element can have variants called isotopes with different numbers of neutrons in the nucleus.

neutron stars The outcome of the collapse of an old star of between 1.4 and 3.2 times the mass of the Sun, primarily consisting of neutrons crammed together resulting in a huge density. A piece of a neutron star the size of a grape would weigh around 100 million tonnes.

positrons Another name for an anti-electron, the positively charged antimatter equivalent of an electron.

quantum numbers The values of quantum states of a particle that can only have integer or half-integer values. An electron in an atom is described by four quantum numbers corresponding to energy level, angular moment, magnetic moment and spin.

spacetime Relativity treats time as a fourth dimension. In relativity there is no absolute position or absolute time because the way things move influences their position in time, so it is necessary to consider spacetime as a whole, rather than to think of space and time independently.

time-reversed waves Maxwell's equations describing electricity and magnetism have two solutions allowing for a wave from the transmitter to the receiver, forwards in time (retarded waves) and from the receiver to the transmitter, backwards in time (advanced waves). Traditionally the time-reversed waves were ignored, but they are useful in explaining mathematical problems with the way an electron recoils when emitting a photon.

wave mechanics An early formulation of quantum theory by Schrödinger that treated particles as 'matter waves'. The wave itself, described by Schrödinger's equation, was interpreted by Max Born as representing probability rather than location. Shown to be equivalent to matrix mechanics.

THE PAULI EXCLUSION PRINCIPLE

the 30-second theory

In 1913 Niels Bohr explained how atoms emit or absorb photons of specific wavelengths when the electrons circling the nucleus leap from one orbit to another in a set of fixed orbits. These orbits are assigned integer quantum numbers (1, 2, 3 ...), called principal quantum numbers. This model worked for hydrogen, the simplest atom known, but in the case of more complex atoms the model did not account for the extra wavelengths that appeared in the spectra of atoms. In 1915 German physicist Arnold Sommerfeld showed that a second quantum number, called the fine structure constant, could account for these wavelengths. In a magnetic field the electrons also behave like tiny magnets, and because they also have spin, physicists had to add a third and a fourth quantum number. The energy of each electron is determined by these four quantum numbers. That same year Wolfgang Pauli discovered that no electrons with the same set of four quantum numbers can orbit the same atom. This concept, the Pauli Exclusion Principle, explains why electrons, even when the atom is in its lowest energy state, distribute themselves over several shells (orbits with the same principal quantum number) and this distribution accounts for the chemical properties of the elements.

RELATED THEORIES
See also
BOHR'S ATOM
page 24

COPENHAGEN INTERPRETATION
page 84

3-SECOND BIOGRAPHIES
NIELS BOHR
1885–1962
Danish pioneer of quantum theory who introduced a first model of the atom

WOLFGANG PAULI
1900–58
Austrian theoretical physicist who introduced the exclusion principle that took his name

30-SECOND TEXT
Alexander Hellemans

The Pauli Exclusion Principle solved the riddle of why the elements with similar chemical properties are arranged in columns in the Periodic Table.

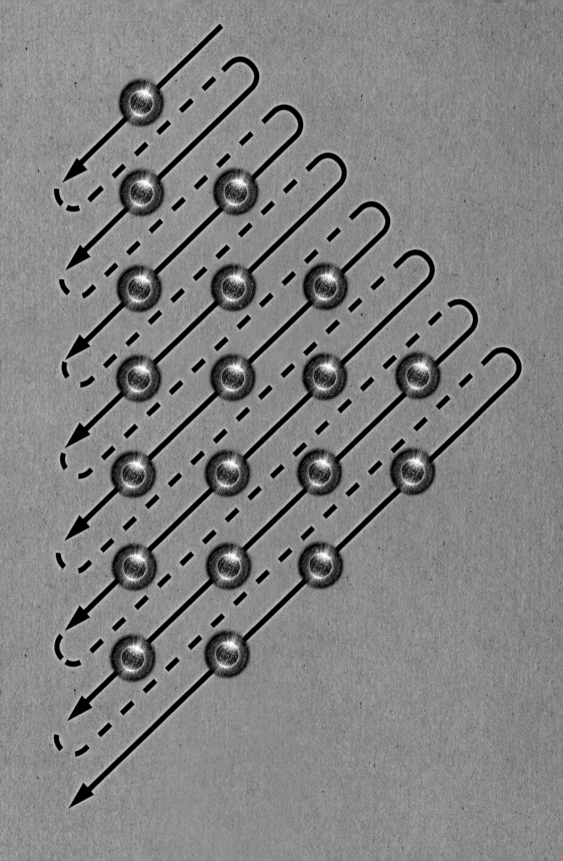

THE DIRAC EQUATION

the 30-second theory

3-SECOND FLASH
In order to arrive at his equation, Dirac had to combine the physics of the very small with the physics of the very fast.

3-MINUTE THOUGHT
The real significance of Dirac's equation may go even deeper than the development of quantum field theory: for a piece of pure mathematics had correctly predicted the existence of a new fundamental particle – Dirac's negative energy electron could equally be an equivalent to an electron with a positive charge. The discovery of a real particle that fitted the description, the positron, identified by Carl Anderson in 1932, may indicate that mathematics is intimately connected to the very fabric of our universe.

Niels Bohr proposed in 1913 that atomic spectra are created when atoms give off and absorb different light wavelengths as electrons jump from one orbit to another. The difficulty was that measurements of the atomic spectra of hydrogen did not fully agree with Bohr's theory. So in the summer of 1927 British theoretical physicist Paul Dirac set about trying to solve this puzzle by analysing the behaviour of electrons. To do this, he welded together the wave equations of quantum mechanics developed by Erwin Schrödinger with the mathematical description of particles moving close to the speed of light embodied in the theory of special relativity. Other physicists had tried this approach but had been stumped by the difficulty of incorporating the fact that electrons possess spin into such a relativistic structure. Dirac resolved this issue through the use of some clever algebra and the incorporation of four-by-four matrices into the equation. The result was a relativistic quantum wave equation, now called the Dirac equation, which had solutions for both positive and negative energy electrons, predicting the existence of antimatter. Dirac's brilliant insight led directly to the development of quantum field theory, the basis of modern particle physics.

RELATED THEORIES
See also
BOHR'S ATOM
page 24

QUANTUM SPIN
page 38

SCHRÖDINGER'S EQUATION
page 42

QUANTUM FIELD THEORY
page 64

3-SECOND BIOGRAPHIES
WILLIAM CLIFFORD
1845–79
British mathematician who first developed the algebra Dirac would later use

CARL ANDERSON
1905–91
Pioneering American experimenter who found anti-electrons in cosmic rays

30-SECOND TEXT
Leon Clifford

Dirac put relativity into the frame to bring theory into line with the observed atomic spectra.

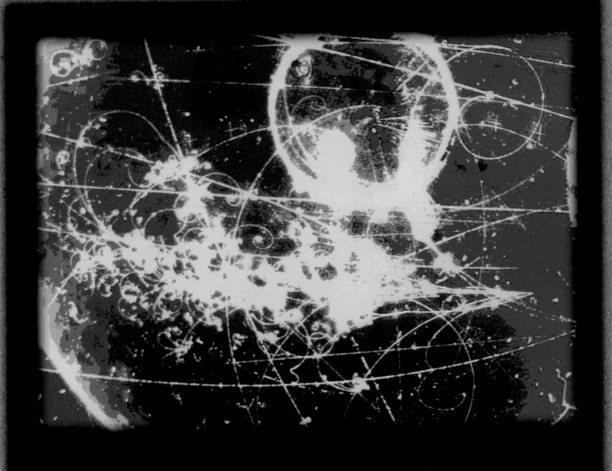

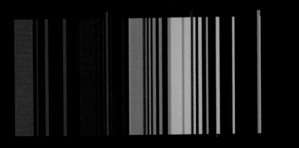

8 August 1902
Born in Bristol to Swiss teacher Charles Dirac and Cornish librarian Florence née Holten

1923
Receives mathematics degree from University of Bristol and begins PhD in Cambridge

1932–69
Lucasian Professor of Mathematics at Cambridge

1921
Receives engineering degree from University of Bristol

1926
Becomes Fellow of St John's College, Cambridge

1933
Wins Nobel Prize in physics (with Schrödinger) for discoveries in atomic theory

1928
Devises the Dirac equation, describing the relativistic motion of an electron

1937
Marries Margit ('Manci') Wigner, sister of physicist Eugene Wigner

1930
Proposes an infinite 'sea' of negative energy electrons and predicts the existence of antimatter

1969
Retires and takes up honorary post at Florida State University

1930
Elected Fellow of the Royal Society

20 October 1984
Dies in Tallahassee, Florida

1932
The antimatter electron, or positron, predicted by Dirac, discovered by Carl Anderson at California Institute of Technology

1952
Received the Copley Medal and the Max Planck Medal

1995
Memorial unveiled in Westminster Abbey

PAUL DIRAC

British physicist Paul Dirac is the greatest scientist to be almost unknown outside his field. Born in Bristol to a Swiss father and a British mother, Dirac had a strict upbringing. It has been suggested that his taciturn nature originated from his father's insistence that Dirac only speak perfect French to him. According to the story, rather than risk getting it wrong, Dirac would not speak at all. But the evidence seems strong that Dirac was on the autistic spectrum, a more likely cause of his lack of interpersonal skills.

Dirac originally studied electrical engineering at the University of Bristol, but his increasing interest in applied mathematics found him taking a second degree before moving on to Cambridge, where he was soon working on relativity and the flourishing new subject of quantum physics. Here Dirac took Schrödinger's already powerful equation, describing the probability of finding a particle in any particular location, and expanded it to take in special relativity for some types of particle, allowing for the effects that high-speed movement would have.

Dirac's equation was symmetrical, allowing for particles that could have either positive or negative energy. This was a serious problem, as an ordinary electron should plunge into the lower negative energy states, pouring out photons. The dramatic solution Dirac proposed was that apparently empty space contained an infinite 'sea' of negative energy electrons, filling all possible negative energy states, preventing electrons from decaying into negative energy. He predicted that this sea could have 'holes' in it – missing negative energy electrons, which would be the equivalent of positive energy anti-electrons, or positrons. He had foreseen the existence of antimatter before it was discovered.

Dirac also contributed a major breakthrough in the development of quantum theory by proving that Heisenberg's matrix mechanics and Schrödinger's wave mechanics, apparently unconnected, were not just consistent but equivalent, pulling the two together to form quantum mechanics.

Dirac shared some personality traits with an earlier holder of his position in Cambridge, Lucasian Professor of Mathematics, Isaac Newton. Like Newton, he had very limited social skills and was infamous for lacking small talk, answering questions as briefly as possible. There are many stories of his attempts at conversation, most notably when meeting the boisterously outgoing American theoretical physicist Richard Feynman. Dirac, after one of his hallmark uncomfortably long pauses, is said to have remarked: 'I have an equation. Do you have one too?' His mathematical brilliance, however, was unquestionable.

Brian Clegg

QUANTUM FIELD THEORY

the 30-second theory

Quantum Field Theory (QFT) is the bedrock of modern particle physics and the mathematical basis of our understanding of the nature of reality. It builds on quantum mechanics by expanding the area of study from a handful of particles to a system of very many particles. It describes the behaviour of fields – physical quantities that have a value at every point in space, a little like the contours on a map – such as the electromagnetic field responsible for light and radio waves, at the quantum level in a way that is impossible to do in quantum mechanics. And, crucially, it enables physicists to deal with both fields and particles at the quantum level within one coherent set of equations. This theory treats waves and particles as if they are disturbances in an underlying field: so, for example, light is a ripple in an electromagnetic field while the electron is an especially excited state of an electromagnetic field. In this way, the theory neatly explains the wave-particle duality found in Nature by combining the wave and particle aspects of both light and electrons – and other combinations of forces and particles – into a single mathematical description of a field.

3-SECOND FLASH
Quantum Field Theory seeks to describe all of the forces and all of the particles found in Nature in terms of the interactions of such fields.

3-MINUTE THOUGHT
So far Quantum Field Theory has been unable to provide a fully consistent quantum description of gravity – the force that operates over the vast distances of space. The successful inclusion of gravity would lead to a unified field theory combining all the known forces and particles that make and shape our world. Such a breakthrough would bring us one step closer to an 'ultimate theory of everything'.

RELATED THEORIES
See also
WAVE-PARTICLE DUALITY
page 28

QUANTUM
CHROMODYNAMICS
page 148

QUANTUM GRAVITY
page 152

3-SECOND BIOGRAPHIES
MARTINUS VELTMAN
1931–
Dutch physicist and one of the pioneers of QFT who helped combine the weak nuclear force with QED

GERARD T'HOOFT
1946–
Dutch physicist who worked with Veltman on the weak nuclear force and worked on QCD and quantum gravity

30-SECOND TEXT
Leon Clifford

The magnetic field that shields the earth from the solar winds requires quantum field theory to describe its quantum mechanical behaviour.

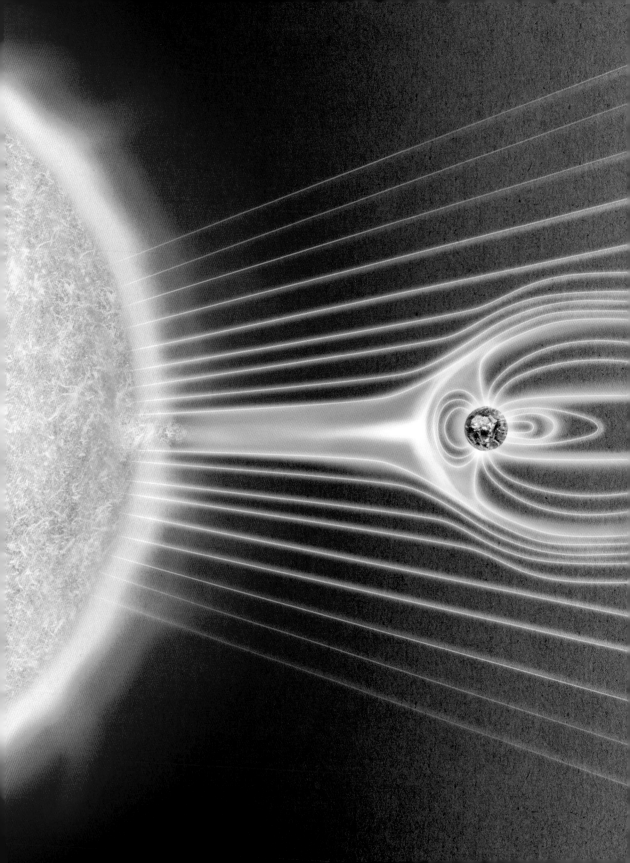

QED BASICS

the 30-second theory

QED – quantum electrodynamics

– takes the classical theory of electromagnetism developed in the 19th century by James Clerk Maxwell into the world of quantum mechanics and special relativity. Classical electromagnetism explains electric currents and electromagnetic waves such as light and radio waves in terms of electromagnetic fields; but the theory was developed before the discoveries of the electron, which carries electric charge, and the photon, which transmits light. Quantum mechanics explains how electrons and photons behave, but is unable to deal effectively with electromagnetic fields. It also has problems with the behaviour of electrons in orbit around atoms where their rapid motion approaches the speed of light and requires the use of the theory of special relativity. QED was inspired by Paul Dirac, who pioneered the successful combination of quantum mechanics and special relativity in his Dirac equation. But the existence of antimatter predicted by Dirac's equation caused a new problem. It implied the possibility of a particle and an antiparticle annihilating themselves in a burst of energy that could condense into many possible combinations of new particles. Dirac realized that this required the development of a new theory that was capable of handling all these particles – and so QED was born.

3-SECOND FLASH
Through the pioneering work of Paul Dirac, QED brought the theory of electromagnetism into the quantum era.

3-MINUTE THOUGHT
The history of physics can in many ways be seen as a series of unifications. Maxwell merged electricity, magnetism and light into a theory of electromagnetism, Einstein unified space and time in special relativity, and quantum mechanics brought together waves and particles. Dirac then welded together special relativity and quantum mechanics and, in turn, these were combined with electromagnetism to form QED. And this story of successive unifications continues right up to the present day.

RELATED THEORIES
See also
THE DIRAC EQUATION
page 60

QUANTUM FIELD THEORY
page 64

QUANTUM CHROMODYNAMICS
page 148

3-SECOND BIOGRAPHIES
JAMES CLERK MAXWELL
1831–79
British scientist who unified electricity, magnetism and light into one consistent theory

PAUL DIRAC
1902–84
British physicist whose work was the inspiration for the development of QED

30-SECOND TEXT
Leon Clifford

Dirac extended Maxwell's classical understanding of electromagnetism to encompass quantum particles.

THE PERILS OF RENORMALIZATION

the 30-second theory

Renormalization is a mathematical technique for solving a problem that arises within quantum field theories such as QED and quantum chromodynamics. This problem involves the appearance of awkward infinities which, without this technique, would make it impossible to tease meaningful solutions from the equations of the theory. Infinities can arise because particle and antiparticle pairs pop into and out of existence for infinitesimally short periods of time within a quantum system. Any attempt to add together the effects of all these different particles quickly leads to infinity, in what mathematicians call a divergent series. In simple terms, renormalization works by parcelling together some of the elements that diverge to infinity and then offsetting them against each other; the remaining balance is replaced in the equation by an arbitrary constant whose value can be determined by experiment. Provided that only a finite number of constants is required and that the values for each such constant can be determined, a theory is said to be renormalizable. For quantum field theories to be accepted by the physics community, they must be shown to be renormalizable. So far, quantum gravity has failed this test.

3-SECOND FLASH

Renormalization is a neat mathematical trick that solves a major problem but was described as 'a shell game' by its most famous inventor, Richard Feynman.

3-MINUTE THOUGHT

Renormalization undoubtedly works, but Richard Feynman was never completely comfortable with the technique he helped to develop. The great physicist Paul Dirac also cautioned against the approach. It is worth remembering that Dirac predicted the existence of antimatter by refusing to ignore some seemingly strange solutions to his own equation. So are these infinities really just mathematical irritations or are they actually telling us something profound about the underlying nature of reality?

RELATED THEORIES
See also
QUANTUM FIELD THEORY
page 64

QED BASICS
page 66

FEYNMAN DIAGRAMS
page 70

3-SECOND BIOGRAPHIES
RICHARD FEYNMAN
1918–88
American physicist and a co-discoverer of renormalization

JULIAN SCHWINGER
1918–94
American physicist credited with renormalizing QED

SIN-ITIRO TOMONAGA
1906–79
Japanese physicist who independently developed renormalization

30-SECOND TEXT
Leon Clifford

Particle/antiparticle pairs popping briefly into existence contribute to the infinities needing renormalization.

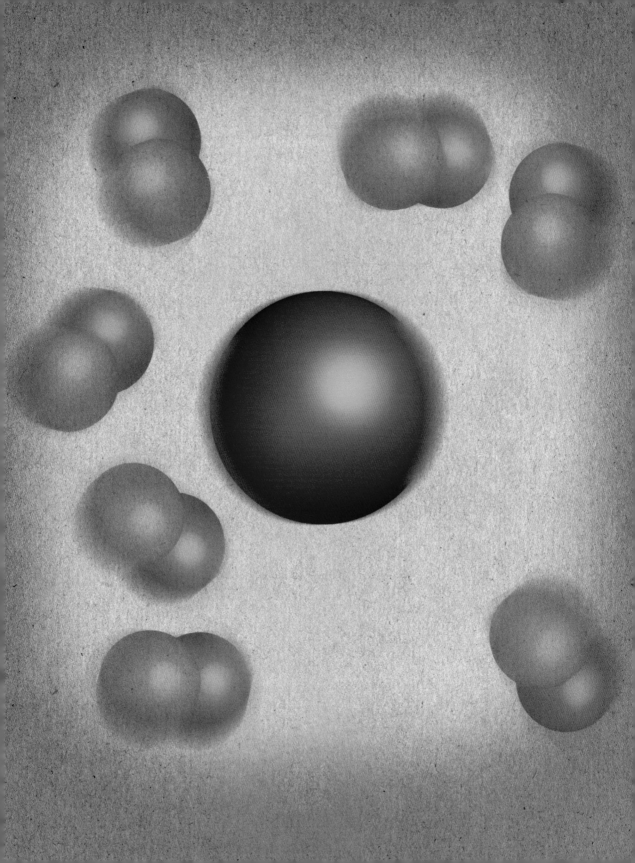

FEYNMAN DIAGRAMS

the 30-second theory

3-SECOND FLASH
Feynman diagrams reduce the world of quantum physics to a more immediate graphic that represents the interaction of particles in time and space.

3-MINUTE THOUGHT
Does the way we view physics affect the way we think about it? Feynman diagrams offer an easily comprehensible visual shorthand that implies a particle-based view of the world. This is at odds with the continuous fields described in quantum field theory (QFT). This reflects the fact that all scientific theories and methods are just models that predict what we observe rather than true descriptions of reality.

A way of visualizing what occurs when a change occurs in the quantum world is to draw a Feynman diagram, a simple graphic that shows how subatomic particles interact. Feynman diagrams were developed in the 1940s by American physicist Richard Feynman as a means of understanding the processes involved in the theory of photons and electrons known as quantum electrodynamics (QED). These diagrams remain relevant to all areas of quantum field theory and have proved hugely successful in helping scientists gain insights into some of the most complex calculations in high-energy particle physics. Every Feynman diagram has to obey a set of specific rules to ensure consistency and usefulness. All Feynman diagrams represent particles by a combination of wavy and straight lines and interactions take place where the lines meet. Feynman diagrams can capture one or more interactions: one axis represents space and another represents time, and lines representing particles move through both space and time diagonally across the diagram. Interestingly, Feynman diagrams show particles of antimatter moving along the time axis in the opposite direction to particles of matter. This can be interpreted as saying that an antiparticle is the equivalent of a particle of normal matter that is going backwards in time.

RELATED THEORIES
See also
QUANTUM FIELD THEORY
page 64

QED BASICS
page 66

THE PERILS OF
RENORMALIZATION
page 68

QUANTUM
CHROMODYNAMICS
page 148

3-SECOND BIOGRAPHY
RICHARD FEYNMAN
1918–88
American physicist, inventor of the eponymous diagrams

30-SECOND TEXT
Leon Clifford

Feynman's elegant diagrams proved an essential tool in understanding quantum electrodynamics.

BACKWARDS IN TIME

the 30-second theory

Waves that travel backwards in time are predicted in the famous equations of electrodynamics developed by James Clerk Maxwell and these predictions are carried into quantum mechanics. According to the mathematics, an event that creates a wave that travels forwards in time – an electron emitting a photon as an electromagnetic wave, say – simultaneously creates a different kind of wave, one that travels backwards in time. These time-reversed waves are known as advanced waves – advanced, since they arrive in advance of their creation. Usually, the bizarre mathematical solutions that allow for advanced waves are ignored, but that does not mean they do not exist. Indeed, one interpretation of quantum mechanics describes quantum events in terms of the interactions of waves going backwards in time with waves that travel forwards in time. However, no one has ever seen an advanced wave. One suggestion is that this is due to the workings of the second law of thermodynamics and that both kinds of waves are actually produced in equal number, as predicted by the mathematics. The action of the second law means that the forward-in-time wave will be absorbed at some point in the future and that this inevitably results in the erasure of all evidence of the advanced waves.

RELATED THEORIES
See also
COPENHAGEN
INTERPRETATION
page 84

MANY WORLDS
INTERPRETATION
page 92

EPR
page 98

3-SECOND FLASH
In quantum mechanics time is a two-way street, where waves travel both backwards and forwards in time, though only the wave travelling forwards is detectable.

3-MINUTE THOUGHT
If advanced waves ever became detectable then it would, in principle, be possible to send a message back in time using an advanced wave by pointing the transmitter at the place in space where the Earth was at some point in the past. What would be the implications of this?

3-SECOND BIOGRAPHIES
JAMES CLERK MAXWELL
1831–79
British scientist who unified electricity, magnetism and light in one consistent theory

JOHN ARCHIBALD WHEELER
1911–2008
American physicist who, with Richard Feynman, explained why it is that we do not see advanced waves

30-SECOND TEXT
Leon Clifford

Unseen advanced waves travel backwards in time to arrive at the source at the very moment of emission.

QUANTUM EFFECTS & INTERPRETATION

alpha particles/alpha decay Alpha particles are helium nuclei consisting of two protons and two neutrons. With beta particles and gamma radiation, they are one of three types of radiation produced in radioactive decay, when the nucleus of an atom loses some of its mass with a release of energy.

Bohm diffusion A mathematical relationship for the rate at which a plasma (a collection of charged ions) diffuses under the influence of a magnetic field. This process is much more complex than the diffusion of a gas, but is described by a simple formula involving only temperature, magnetic field strength and a constant.

Manhattan Project The Allied Second World War project to make an atomic bomb, set up in response to intelligence that Germany was attempting to build one. Led by and located in the United States, but with significant contributions from the UK and Canada. Named after the temporary headquarters of the Army faction, based on Broadway, the project pulled together many sites, though the eventual focus was on Los Alamos in New Mexico. The first bomb test, codenamed Trinity, took place at what is now the White Sands Proving Ground on 16 July 1945, less than four weeks before the bomb's deployment in Japan.

photon A quantum particle of light and the carrier of the electromagnetic force. Until the 20th century light was thought to be a wave, but both theory and experiment showed that it can also be treated as a massless particle.

superposition Superposition is a fundamental behaviour of quantum theory that has no equivalent in the macro world of objects we see around us. It says that where a quantum particle has a state that, say, has two possible values – such as spin, which can be 'up' or 'down' – it will not have an actual value but simply have a probability of being in one state or the other, until it is measured when it collapses to a single, actual value. A tossed coin is a real world item with two states. Before we look at the coin it could be heads or tails with 50 per cent probability – but we know that it actually has one of these values. One side is face up. A quantum particle, though, has no value, just the probabilities while it is in superposition.

thought experiment An experiment that is not actually carried out, but that can be used to demonstrate a concept or an idea. Schrödinger's Cat (see page 46) is probably the best-known thought experiment in physics, but Einstein, for example, was always using them, particularly in his attempts to discredit quantum theory. His best-known, the EPR thought experiment (see page 98), led to the development of actual experiments demonstrating quantum entanglement.

wavefunction In quantum physics, the wavefunction is a mathematical formula that describes the behaviour of a quantum state of the particle, which evolves over time according to the Schrödinger wave equation. The wave in question, which spreads out over time, does not describe the particle itself, but rather the probability of a quantum state having a particular value – so, for instance, it can describe the probability of finding a particle in different locations. The probability is given by the square of the wavefunction.

zero time tunnelling Because a quantum particle does not have an exact location before it is observed, it can pass through an obstacle that it should not be able to get past. This process is known as quantum tunnelling. In experiments in which a particle is measured passing along a route that includes a barrier, the particles tunnelling through the barrier appear to spend no time inside the barrier, hence 'zero time tunnelling'.

BEAM SPLITTERS

the 30-second theory

Everyone has experienced a sophisticated quantum device called a beam splitter, because a glass window is a great example. Stand in a room at night with the light on and look out of the window. You can see yourself reflected clearly. But go outside and you can also see into the room. While some light from the room reflects back in – maybe 5 per cent – most passes through. (It occurs all the time, but is only obvious at night, as the reflection is barely visible in daylight.) This presents an interesting problem. Newton thought light was made of particles, but couldn't explain why a particular particle reflected or passed through glass. He thought it might be caused by imperfections in the surface, although this isn't supported experimentally. We now know it is due to the quantum nature of photons. We can't tell if a photon will reflect, merely the probability. The effect is even more remarkable because the percentage reflected from the inner surface depends on the thickness of the glass. The incoming photons are influenced by this because as quantum particles they are sufficiently spread out to interact with the whole sheet, not just the surface.

3-SECOND FLASH
A window is a quantum device called a beam splitter that allows a percentage of photons through. This baffled Newton, unaware of the probabilistic nature of quantum particles.

3-MINUTE THOUGHT
Beam splitters can entangle particles, even collections of particles. The process starts with a photon sent unobserved through a beam splitter. It is in a superposition of states – we only have probabilities of whether it reflected off or through. Each path interacts with a cloud of atoms 'say' then passes through a polarizing beam splitter, choosing direction on polarization. Finally, the photon is detected, triggering the entanglement of the atom clouds. The detail involves messy maths – but the process works.

RELATED THEORIES
See also
QUANTUM DOUBLE SLIT
page 32

QUANTUM TUNNELLING
page 80

EPR
page 98

3-SECOND BIOGRAPHIES
ISAAC NEWTON
1642–1727
British physicist most famous for gravitation and laws of motion, but also very active in optics

MICHAEL HORNE
1943–
Leading American physicist in quantum theory and entanglement, another beam splitter expert

30-SECOND TEXT
Brian Clegg

Most light passes through window glass, but a proportion is reflected, a process Newton struggled to explain.

QUANTUM TUNNELLING

the 30-second theory

If a ball rolls up a hill but lacks the energy to get to the top, it will never reach the other side. This seems obvious, but in quantum physics it isn't true. A quantum object such as an electron or a photon can pass through a barrier even if in classical terms it doesn't have enough energy. This so-called quantum tunnelling is a consequence of quantum particles not having a well-defined location, just the wavefunction that describes the probability of finding the object at different points in space. The presence of a barrier attenuates the wavefunction but doesn't shrink it to nothing even on the far side: there is a finite, if small, chance that the object might be found there. Tunnelling plays an important role in several natural phenomena. It is what enables alpha particles to escape from the strong binding forces of an atomic nucleus in radioactive decay; it can speed up the rate of some chemical processes and allows some chemical reactions to proceed in cold interstellar space. The effect is used technologically, for example in certain kinds of diode in which electrons tunnel across the junction between two types of semiconductor.

RELATED THEORIES
See also
JOSEPHSON JUNCTIONS
page 126

3-SECOND BIOGRAPHIES
FRIEDRICH HUND
1896–1997
German physicist, a pioneer of quantum chemistry who first recognized the importance of tunnelling in the light-emission spectra of molecules

GEORGE GAMOW
1904–68
Russian-born American theoretical physicist who recognized the importance of quantum tunnelling in alpha decay

GERD BINNIG & HEINRICH ROHRER
1947– & 1933–2013
German and Swiss inventors of the scanning tunnelling microscope in the 1980s, for which they received the 1986 Nobel Prize in Physics

30-SECOND TEXT
Philip Ball

3-SECOND FLASH
Quantum tunnelling is the penetration of quantum particles through a barrier even though in classical terms the particles lack enough energy to surmount it.

3-MINUTE THOUGHT
Quantum tunnelling can be useful in semiconductor microelectronics, but it is also a nuisance. As transistors on silicon chips get smaller, the insulating layers separating components become thinner, just a few atoms across. This makes them leaky barriers, since electrons may tunnel through them. Then it becomes impossible to turn the devices off. So far the problem has been solved by replacing the silicon dioxide insulator with a better one made of hafnium dioxide.

A tunnelling particle gets around a barrier by moving from A to B without going through the space in between.

Waterstones

St John's Centre
Perth
PH1 5UX
01738 630013

SALE TRANSACTION

30-SECOND QUANTUM T   £14.99
9781848316669
CARRIER BAG - SCOIL   £0.05
2000069007882

Balance to Pay   **£15.04**
Gift Card Tendered   £15.00
Cash   £0.05
CHANGE   £0.01

- - - - - - - - - - - - - - - - - - -

WATERSTONES REWARDS POINTS

WATERSTONES REWARDS POINTS WOULD HAVE
EARNED YOU 45 POINTS TODAY
ON ITEMS WORTH £15.04
Apply now at thewaterstonescard.com

- - - - - - - - - - - - - - - - - - -

Fantastic Beasts and Where to Find Them
The Original Screenplay by JK Rowling
Pre-order your copy at half price now

VAT Reg No. GB 108 2770 24

9992035600218244I4

manufacturer.

This does not affect your statutory rights.

Waterstones Booksellers,
203/206 Piccadilly, London, W1J 9HD.

Get in touch with us:
customerservice@waterstones.com
or 0808 118 8787.

Buy online at Waterstones.com or Click & Collect.
Reserve online. Collect at your local bookshop.

Did you love the last book you read? Share your
thoughts by reviewing on Waterstones.com

Waterstones

Refunds & exchanges

We will happily refund or exchange
goods within 30 days or at the manager's
discretion. Please bring them back with
this receipt and in resalable condition.
There are some exclusions such as Book
Tokens and specially ordered items, so
please ask a bookseller for details.

If your Kindle is faulty we will replace it
when returned within 30 days. For those
returned after 30 days but within the
manufacturer's warranty period, we will
gladly arrange a replacement from the
manufacturer.

This does not affect your statutory rights.

Waterstones Booksellers,
203/206 Piccadilly, London, W1J 9HD.

Get in touch with us:
customerservice@waterstones.com

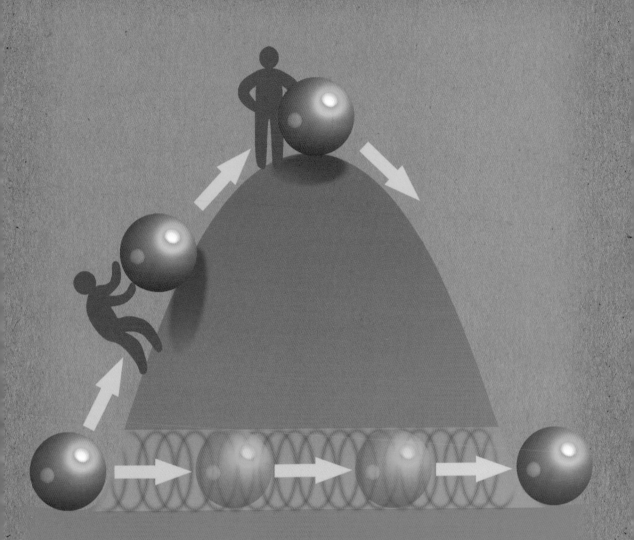

SUPERLUMINAL EXPERIMENTS

the 30-second theory

A remarkable outcome of quantum physics is that photons can travel faster than the speed of light. Such 'superluminal' experiments send photons towards a barrier through which they should not pass. Quantum theory tells us that a photon's location is not fixed and there is a small probability that it will already be at the other side of the barrier. So a few photons instantly pass through it and carry on. If the barrier is 1 unit of distance wide, and the photon travels the same distance either side of the barrier, it traverses 3 units of distance in the time light usually takes to cover 2 units, moving at 1.5 times the speed of light. Early experimenter Raymond Chiao insists that it is impossible to send a signal this way, because photons that get through a barrier are random. However in 1995, Günter Nimtz modulated the tunnelling beam, demonstrating this by playing a recording of Mozart's Symphony No. 40 transmitted at over four times the speed of light. There remains dispute over whether the signal truly travels faster than light or is simply distorted by the process, rather like a runner leaning forward to break the tape first.

RELATED THEORIES

See also
SCHRÖDINGER'S EQUATION
page 42

QUANTUM TUNNELLING
page 80

JOSEPHSON JUNCTIONS
page 126

3-SECOND FLASH
Because quantum particles that tunnel through a barrier do so in zero time, photons undergoing tunnelling appear to travel faster than light.

3-MINUTE THOUGHT
Most early experiments used highly technical barriers called undersized waveguides or photonic lattices. Nimtz often uses an example of tunnelling discovered by Newton – frustrated total internal reflection. When a beam of light enters a prism at a right angle it bounces off the back of the glass. Newton discovered that a second prism, placed close but not touching, enables part of the beam to flow through instead of reflecting. This happens because photons tunnel through the barrier formed by the gap.

3-SECOND BIOGRAPHIES
ISAAC NEWTON
1642–1727
British physicist famous for gravitation and laws of motion, but also very active in optics

GÜNTER NIMTZ
1936–
German physicist who has worked on the impact of electromagnetic radiation on humans and superluminal effects of quantum tunnelling

RAYMOND CHIAO
1940–
American physicist who specializes in quantum optics

30-SECOND TEXT
Brian Clegg

Nimtz used tunnelling between two prisms to send a Mozart symphony faster than light.

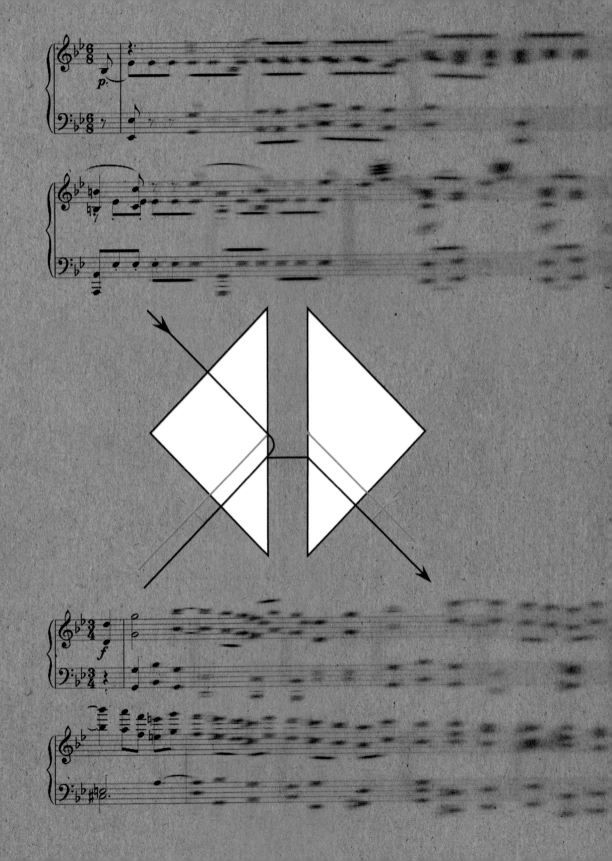

COPENHAGEN INTERPRETATION

the 30-second theory

3-SECOND FLASH
The Copenhagen Interpretation states that it is not meaningful to think of anything more fundamental to quantum systems than what we are able to measure.

3-MINUTE THOUGHT
A central component of the Copenhagen Interpretation is Bohr's notion of complementarity. This means you can answer some questions with one experiment, and others with another but they won't necessarily give consistent results. In one you see a particle; in another a wave. Neither is more true: both are needed. Although many physicists accept this, they are less enthusiastic now about the readiness of Bohr and his followers to extend complementarity to biology, ethics, religion, politics and psychology.

As quantum theory took shape in the 1920s, it looked ever stranger. The Schrödinger equation implied that particles could act like waves. Quantum particles could exist in superpositions. Heisenberg formulated his uncertainty principle. What did it all mean? Niels Bohr, working in Copenhagen and assisted by Heisenberg and others, provided a disconcerting analysis, now known as the Copenhagen Interpretation, which broke with the longstanding belief that science can always find things out. It accepts that in quantum theory there are unanswerable questions, and that one experiment might not be consistent with another. All we can know about the world is what we can measure. Take the classic experiment of a beam of photons being fired through two parallel slits. If you don't ask which slit the photons pass through, they act like waves forming an interference pattern of bright and dark on the far side. If you set up the apparatus to detect which slit they go through, there's no interference pattern. But which slit did they 'really' go through to cause interference? According to the Copenhagen Interpretation you can't ask that. There is no essential reality beyond the quantum description, nothing more fundamental than probabilities and measurements.

RELATED THEORIES
See also
QUANTUM DOUBLE SLIT
page 32

WAVE-PARTICLE DUALITY
page 28

SCHRÖDINGER'S EQUATION
page 42

HEISENBERG'S UNCERTAINTY PRINCIPLE
page 48

3-SECOND BIOGRAPHY
DAVID MERMIN
1935–
American physicist who famously paraphrased the Copenhagen interpretation as 'Shut up and calculate!'

30-SECOND TEXT
Philip Ball

Complementarity says that light can act as waves (top) or as particles (bottom) but not both simultaneously.

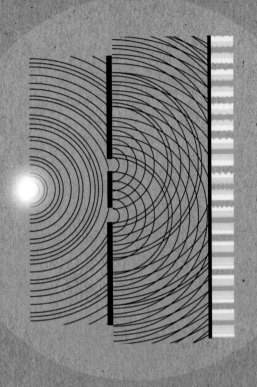

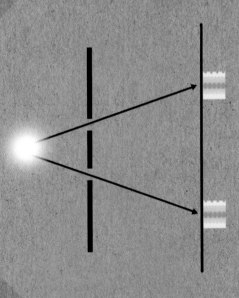

BOHM INTERPRETATION

the 30-second theory

3-SECOND FLASH
Bohm attempted to remove the element of chance from quantum mechanics, challenging the widely accepted theory.

3-MINUTE THOUGHT
If reality is ultimately based on some deterministic framework, such that every aspect of the world that we inhabit, including the workings of our brains, is determined by the physical laws governing the whole of the universe, then does free will really exist?

There is an alternative to the Copenhagen Interpretation of what happens when an attempt is made to measure a quantum system, causing the wave function to collapse. It states that there is no such measurement problem in quantum physics since particles are only ever in one place at any time even when no one is around to look at them. For example, in the double-slit experiment, particles do not pass through both slits simultaneously, instead each particle only passes through one slit. In this model, the wave function serves to determine the distribution of particles at the end of the experiment and the apparent localized collapse of the wave function is just the result of making a particular measurement at a particular time on discrete particles that were already following clearly determined paths. Such a causal and deterministic approach to quantum mechanics is in stark contrast to the more widely accepted probabilistic treatment. This radically different view is known as the Bohm Interpretation after David Bohm, the American-born theoretical physicist who developed it. Similar ideas were proposed in the early days of quantum mechanics by Louis de Broglie and the explanation is sometimes called the de Broglie-Bohm theory.

RELATED THEORIES
See also
COPENHAGEN INTERPRETATION
page 84

MANY WORLDS INTERPRETATION
page 92

EPR
page 98

BELL'S INEQUALITY
page 100

3-SECOND BIOGRAPHY
LOUIS DE BROGLIE
1892–1987
French physicist who originally pioneered a causal approach to quantum mechanics

30-SECOND TEXT
Leon Clifford

Bohm's interpretation would restore a clockwork universe where everything is predetermined.

20 December 1917
Born in Wilkes-Barre, Pennsylvania, United States

1939
Awarded BSc from Pennsylvania State College

1940
Joins Robert Oppenheimer at University of California, Berkeley, as postgraduate research student

1943
Awarded PhD on the basis of research into nuclear scattering. Works at Berkeley Radiation Laboratory, contributing calculations for the Manhattan Project

1947
Moves to become Assistant Professor of Physics at Princeton, where he works alongside Albert Einstein and conducts research into plasmas, metals and quantum mechanics

1949
Discovers a law governing how plasmas diffuse in magnetic fields – now called Bohm diffusion– and publishes a paper describing the phenomenon

1951
Moves to Brazil and publishes his first book, *Quantum Theory*, taking a conventional view

1957
Moves to the UK and publishes *Causality and Chance in Modern Physics*, which expounds his deterministic view of quantum mechanics

1959
With Yakir Aharonov, discovers the Aharonov-Bohm Effect showing that electromagnetic potentials are real, not simply mathematical concepts

1980
Publishes *Wholeness and the Implicate Order*, outlining his belief that there is a deeper underlying basis for reality

1990
Elected a Fellow of the Royal Society

27 October 1992
Dies in London

1993
The Undivided Universe: An Ontological Interpretation of Quantum Theory, a key text explaining the Bohm Interpretation, co-written with Basil Hiley, published posthumously

A quest for deeper order in the

world defined the life of theoretical physicist David Bohm who cast doubt on the Copenhagen Interpretation. This search inspired his work in quantum mechanics, caused him to flirt with Eastern mysticism in his later years and led him during the 1930s towards communism; an affiliation that ultimately required him to leave the United States, the country of his birth, and live in effective exile.

In 1949 Bohm refused to testify to Congress against his former PhD supervisor and suspected communist sympathiser, Robert Oppenheimer. Bohm was arrested and charged with contempt of Congress. He was later tried and acquitted, but the scandal led to him being fired from his job at Princeton. After this, Bohm worked abroad, moving first to Brazil in 1951, then to Israel in 1955. He finally settled in the UK in 1957 and became Professor of Theoretical Physics in 1961 at Birkbeck College, University of London, where he developed the detail of his interpretation of quantum theory.

Two great intellectual friendships influenced Bohm: the physicist Albert Einstein at Princeton and the philosopher Jiddu Krishnamurti in London; both men in their different ways aided Bohm as he sought out order in science and society. Einstein's nagging doubts about quantum mechanics and his view that 'God does not play dice' undoubtedly struck a chord with the young Bohm, while Krishnamurti's spiritual viewpoint helped Bohm put his idea of the oneness of the universe into a philosophical context.

Bohm came to believe that there is some deeper reality to the universe and that the world we see around us is akin to a ghost, a projection of this hidden truth. For Bohm, true reality could only be glimpsed by a mind that was free from the self-deceptions created by the very process of thinking. His interpretation was that the universe we see – the universe of space and time and particles and quantum mechanics – unfolds naturally out of this deeper underlying reality, what he called the implicate order.

Bohm's belief resulted in a new interpretation of quantum mechanics, one that invokes a wave function for the whole universe, that can evolve according to the Schrödinger equation and that is deterministic, guiding the path of every particle in existence. This causal and deterministic treatment contrasts with the probabilistic explanation of the Copenhagen Interpretation. Bohm's unorthodox view of quantum mechanics has never been widely accepted by physicists, but remains a coherent alternative interpretation.

Leon Clifford

CONSCIOUSNESS COLLAPSE

the 30-second theory

Quantum wavefunctions

collapse when attempts are made to observe and measure quantum systems. When this occurs, all possible states of the quantum system coalesce into the one observed state, a phenomenon that has given rise to the Copenhagen, Many Worlds and Bohm interpretations of quantum mechanics. But the questions about what causes the wavefunction to collapse and at what point in the measurement process the actual collapse occurs remain subjects for debate. One suggestion (no longer widely held) was that the wavefunction only collapses when a conscious observer is involved in the measurement. Conscious observers can only see the world in one way, therefore they must be in one state or another and cannot be simultaneously in many states; it is this requirement of consciousness to be in a single state that forces the wavefunction to collapse. To explain the idea, physicist Eugene Wigner proposed a version of the famous Schrödinger's Cat thought experiment in which a friend was placed in the box with the cat. Wigner suggested that the presence of the conscious mind of the friend would cause the wavefunction to collapse inside the box, crystallizing the state of the cat as either alive or dead.

3-SECOND FLASH
It seems that you can affect the quantum world just by looking at it – but does this require consciousness?

3-MINUTE THOUGHT
The possibility that our minds somehow interact with the quantum world raises the question of whether our consciousness may itself be a quantum phenomenon. After all, our brains are made from atoms and use electrical signals, which are all subject to the laws of physics. Could quantum mechanics one day provide an explanation for the mystery of human consciousness?

RELATED THEORIES
See also
SCHRÖDINGER'S CAT
page 46

COLLAPSING
WAVEFUNCTIONS
page 50

3-SECOND BIOGRAPHIES
EUGENE WIGNER
1902–95
Hungarian-born physicist who first proposed that collapse of the wavefunction occurred as a result of interacting with our consciousness

JOHN VON NEUMANN
1903–57
Hungarian-born mathematician, who saw consciousness as a part of the chain involved in the collapsing quantum wavefunction

30-SECOND TEXT
Leon Clifford

Some physicists have suggested that it takes a conscious observer, like a human being, to cause a wavefunction to collapse.

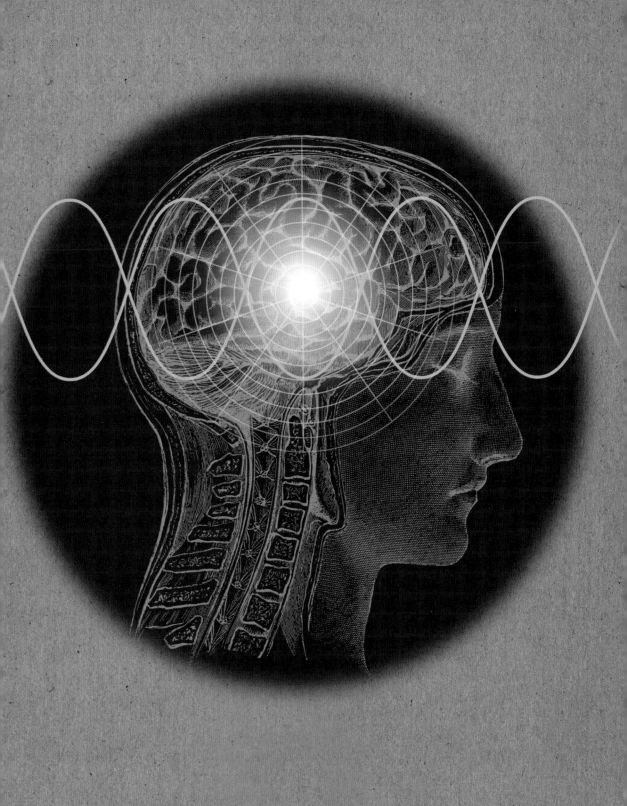

MANY WORLDS INTERPRETATION

the 30-second theory

3-SECOND FLASH
Originally called relative state formulation, the 1957 theory now known as Many Worlds proposes the existence of an infinite set of parallel universes.

3-MINUTE THOUGHT
If Many Worlds is true then that most famous paradox of quantum theory, Schrödinger's Cat, is no longer a problem. In one world the cat is alive, in another it is dead. It can not be both simultaneously in the same world. All possible outcomes of each quantum decision exist in their own branches of reality. However, many physicists believe the extra complexity of Many Worlds, requiring a new world each time every particle in the universe undergoes a change, is too high a price to pay to get around the strangeness of the Copenhagen Interpretation.

Many physicists accept the Copenhagen Interpretation of quantum physics that quantum particles genuinely can be in more than one state simultaneously, and that the probability wave that predicts their position enables them to act as if they were in more than one place. For some, though, this is a step too far. Hugh Everett was determined to find a way of rationalizing the strange behaviour of quantum particles. In his controversial PhD thesis he presented the theory that would dominate his working life, the Many Worlds Interpretation. It dispenses with the idea of waveforms collapsing to provide a specific value on being observed. Instead, according to Many Worlds, each time a quantum particle can have more than one state, the world branches. The particle exists in one state in one version of the universe and in the other in the second. The reality we experience is just a single path through each of these worlds. This means we no longer need worry how a photon or electron somehow interferes with itself when passing through a double-slit experiment – in one universe it goes through the first slit, in another it goes through the second. Although we only experience a single universe directly, we can see the result of the different universes interacting in the interference pattern of light and dark fringes produced by the double slits.

RELATED THEORIES
See also
SCHRODINGER'S CAT
page 46

COPENHAGEN INTERPRETATION
page 84

BOHM INTERPRETATION
page 86

3-SECOND BIOGRAPHIES
HUGH EVERETT III
1930–1982
American physicist who first proposed the interpretation

BRYCE DEWITT
1923–2004
American physicist who popularized and named the Many Worlds Interpretation

30-SECOND TEXT
Brian Clegg

In Many Worlds, whenever there is more than one possible outcome at the quantum level, there is a different world to accommodate it.

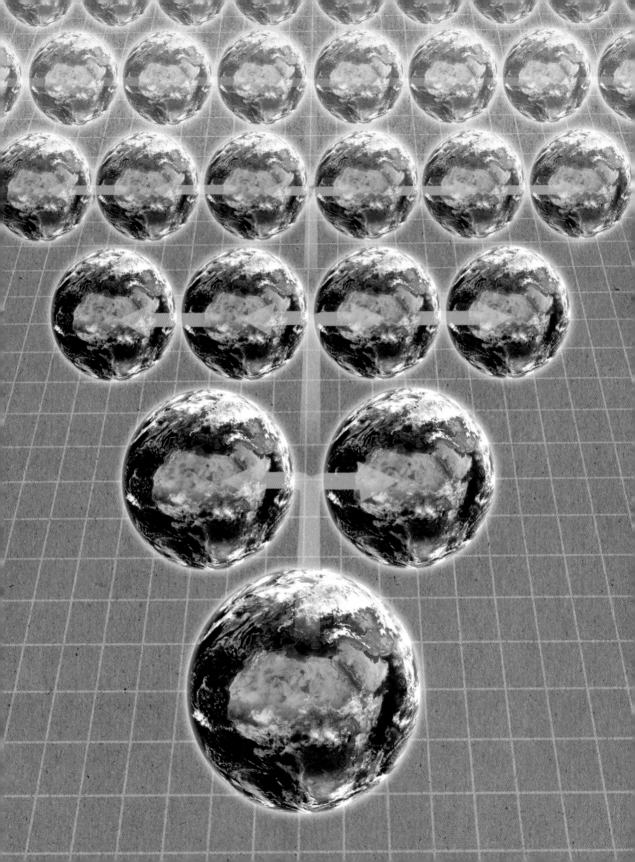

QUANTUM ENTANGLEMENT

bit A contraction of BInary digiT. The basic unit of storage in computing, a bit can have a value of 0 or 1.

conservation of momentum The momentum of an object is its mass times its velocity. Momentum is conserved, so if for instance a stationary particle (with zero momentum) splits into two moving particles, those particles must have equal and opposite momentum.

encryption The use of codes and cyphers to conceal the meaning of information.

hidden variables Einstein and others doubted the probability-based nature of quantum theory. They believed that there was an underlying reality that provided the actual value observed, rather than its being dependent on probability. The concealed values are called hidden variables.

local reality The idea that a quantum particle can only influence another particle if it is nearby (locality) and if its properties have real values (reality). Einstein's EPR thought experiment (see page 98) established that either quantum theory was flawed or local reality did not exist for quantum particles.

microwave cavity A metal chamber holding an electromagnetic wave in the microwave part of the spectrum. The chamber acts as a resonator, just as a string can vibrate with a particular frequency, but rather than a physical wave fitting between two fixed points on a string, in a microwave cavity there is an electromagnetic wave fitting between two 'fixed' points created by the metal walls of the chamber.

MRI scanner A medical device, formerly known as a NMR (nuclear magnetic resonance) scanner, that uses powerful, often superconducting magnets to manipulate the quantum spin of protons in the nucleus of the hydrogen in water molecules, usually in living things. When the molecules flip they act as tiny transmitters, whose output is then detected. See page 124.

one-time pad An unbreakable means of encryption devised in 1918. Each character in the message to be encrypted has a randomly selected value added to it. The final message is itself random and not susceptible to cracking. Despite being unbreakable, it was not widely used because the list of values (the 'pad') must be provided to both ends of the communication and this pad can be intercepted.

quantum dots Nanoparticles of a semiconductor that act as an artificial atom. They are used in quantum technology, notably in electronics and solar cells, and as qubits in quantum computers. See page 130.

quantum entanglement A fundamental aspect of quantum theory: two (or more) quantum particles can be linked together in such a way that a change made to the state of one particle is reflected instantly in the other, however far apart they are separated. Einstein believed that this was impossible, as the particles 'communicate' faster than the speed of light, but it has been repeatedly demonstrated in experiment.

qubit The quantum computing equivalent of a bit. Where a bit can only have a value of 0 or 1, a qubit can be in a superposition of states where it represents the probabilities of being 0 or 1. The qubits can also be entangled, multiplying up the combinations of values, so that a much larger computation can be done with qubits than could be done with the same number of bits.

superposition When a quantum particle has a state with, say, two possible values it will not have an actual value but rather a superposition – a collection of probabilities of being in the states – until it is measured, when it collapses to an actual value. A tossed coin has two states but no superposition. Before we look, the coin already has one of these values. But a quantum particle has no value, just probabilities, while in superposition.

EPR

the 30-second theory

In 1935, Albert Einstein joined with younger colleagues Boris Podolsky and Nathan Rosen to write an academic paper he hoped would prove his long-held belief that quantum theory was incorrect. Formally entitled *Can Quantum-Mechanical Description of Physical Reality Be Considered Complete?* it is universally known by its authors' initials, 'EPR'. The paper imagines a particle splitting into two, with the new particles flying off in opposite directions. According to quantum theory, after a while the momentum of the particles does not have an absolute value, merely a range of probabilities. Measure the momentum of one particle and it takes a specific value. Instantly, however far apart the particles are, the other particle's momentum must become equal and opposite to fulfil conservation of momentum. A similar effect can be obtained on measurements of position. The EPR paper concludes that either quantum theory is wrong, and there are hidden values that specify the momentum (and so on) before measurement, or locality – the idea that one thing cannot influence another at a distance without something passing between them – has to be thrown out. Einstein and friends conclude 'No reasonable definition of reality could be expected to permit this.' Here, Einstein thought, was the clincher. Experiments would prove him wrong.

3-SECOND BIOGRAPHIES
ALBERT EINSTEIN
1879–1955
German-born physicist who, though sceptical, contributed to quantum theory

BORIS PODOLSKY
1896–1966
American physicist who may have worded EPR

NATHAN ROSEN
1909–95
American-Israeli scientist who worked on EPR and later devised the idea of the wormhole

30-SECOND TEXT
Brian Clegg

The spooky linkage of entanglement means that observing a value in one particle instantly influences the other.

3-SECOND FLASH
Einstein's EPR thought experiment, designed to destroy quantum theory, demonstrated when put into practice that Einstein was mistaken.

3-MINUTE THOUGHT
EPR can be a little confusing as it refers to both the momentum and position of the particle, which is reminiscent of the uncertainty principle. But EPR would have worked just as well with a single property. When Schrödinger pointed out how using two properties hid the meaning of the paper, Einstein replied that using two properties '*ist mir Wurst*', literally 'is sausage to me', colloquial German for: 'I couldn't care less.'

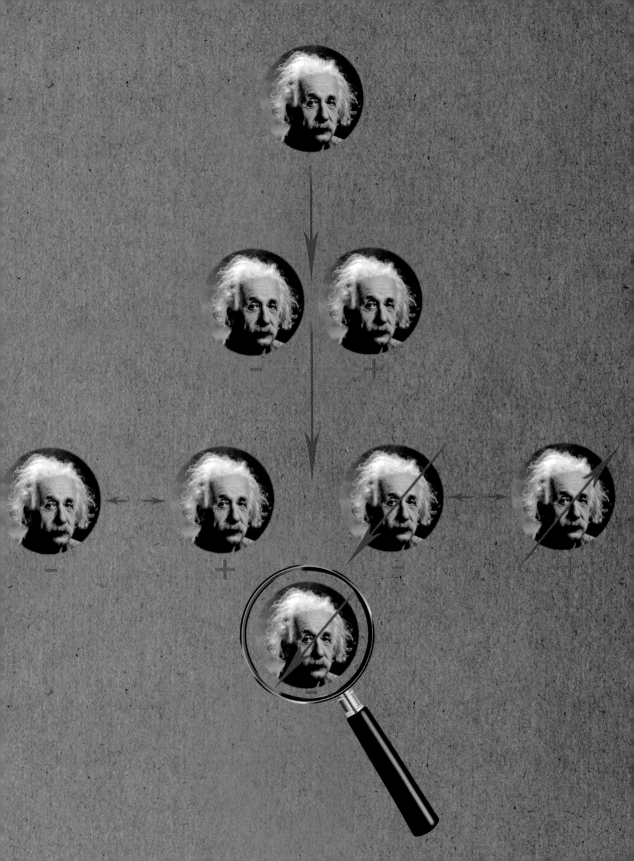

BELL'S INEQUALITY

the 30-second theory

3-SECOND FLASH

Einstein disagreed with Schrödinger's idea of quantum entanglement, which he called 'spooky action at a distance', but did not live to see the experiments that vindicated Schrödinger.

3-MINUTE THOUGHT

Two entangled particles can be viewed as a single physical object, even if they are light years apart. Quantum entanglement will be a powerful tool in future computing and in data encryption. While bits in current computers are switched with electric pulses, qubits will be linked by entangling them with subatomic particles.

The basic idea of Schrödinger's Cat is the superposition of quantum states: both the atomic nucleus and the cat in the box are in two states simultaneously. If you open the box, you find the cat either dead or alive, and the nucleus decayed or intact. In quantum-speak, the cat and the atomic nucleus are 'entangled'. Typically, two identical particles created in one process are entangled, and remain so even if they become separated by long distances. Both these particles are in a superposition of two quantum states, but if you check on one of them, your measurement immediately affects the quantum state of the other particle. Einstein and colleagues Podolsky and Rosen argued that if two particles remain entangled over long distances, a physical influence between them has to travel faster than light, which contradicts the theory of relativity. In 1964, John Bell produced a measurement that enabled experimenters to distinguish between a link that took place at the moment of measurement and one in which there were 'hidden variables' that set up the values that would be measured before the particles separated. This distinguishing factor was Bell's inequality. In 1984, Alain Aspect performed such an experiment on photons, supplying experimental proof of their entanglement.

RELATED THEORIES

See also
SCHRÖDINGER'S CAT
page 46

COPENHAGEN
INTERPRETATION
page 84

BOHM INTERPRETATION
page 86

MANY WORDS
INTERPRETATION
page 92

QUANTUM COMPUTING
page 108

3-SECOND BIOGRAPHY
ALAIN ASPECT
1947–
French experimental physicist who demonstrated quantum entanglement

30-SECOND TEXT
Alexander Hellemans

The endangered cat in Schrödinger's thought experiment is entangled with the nucleus of a decaying atom.

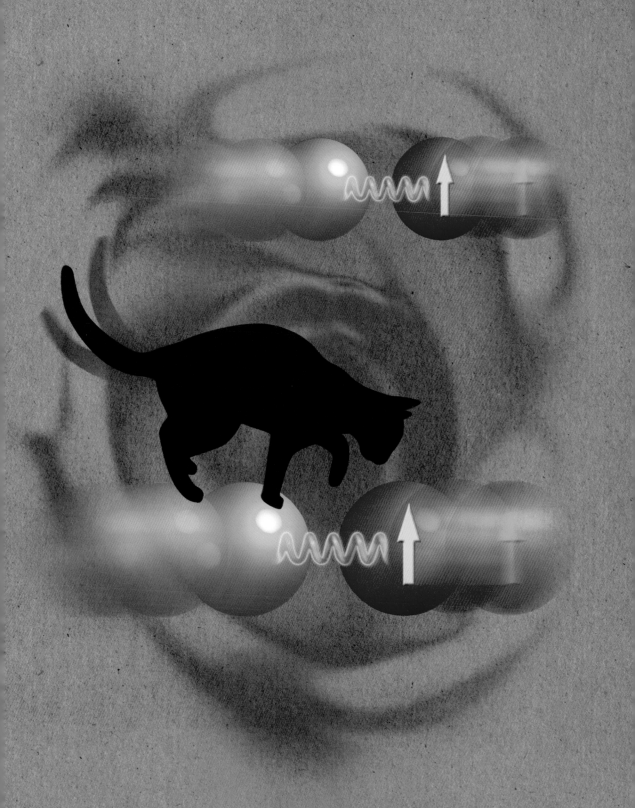

28 July 1928
Born in Belfast to John Bell and Annie née Brownlee

1948
Receives experimental physics degree from Queen's University, Belfast

1949
Receives mathematical physics degree from Queen's and goes to work at the UK Atomic Research Establishment, Harwell, Oxfordshire

1954
Marries Mary Ross

1956
Completes PhD in physics at University of Birmingham

1960
The Bells move to work at CERN near Geneva

1964
Publishes breakthrough paper, 'On the Einstein-Podolsky-Rosen Paradox', specifying Bell's inequality

1972
American group of John Clauser, Abner Shimony, Michael Horne and Richard Holt provides first experimental verification of Bell's theorem, supporting quantum theory, but their approach has a possible loophole

1982
French physicist Alain Aspect closes the loophole, vindicating quantum theory using Bell's theorem

1987
Elected Foreign Honorary Member of the American Academy of Arts and Sciences

1 October 1990
Dies in Geneva, Switzerland

2008
The John Stewart Bell Prize for research in fundamental issues in quantum mechanics is created

JOHN BELL

John Bell's brothers and sister

left school at 14, so it was a surprise when young Stewart (the family referred to him by his middle name to distinguish him from his father) announced he intended to go to university and become a scientist. Although this was something new for the Bell family, his mother encouraged it, wanting 'the Prof', as they sometimes called him, to have a life 'where he could wear his Sunday suit all week!'

Bell attended Belfast Technical High School and Queen's University, Belfast. Afterwards, rather than stay in academia, the financially conscious Bell got a job with the UK Atomic Research Establishment at Harwell in England. While at Harwell he met his wife-to-be, Mary Ross, a Scottish physicist, and the pair secured posts at CERN, the European Centre for Nuclear Research near Geneva.

Although particle physics was Bell's bread and butter job, a sabbatical in 1963 gave him a chance really to think about quantum theory, a subject that had always fascinated him. Bell had some sympathy with Einstein's view that there was something uncomfortable about quantum theory, that there ought to be reality underlying the apparent randomness. He once said of quantum physics: 'I hesitated to think it might be wrong, but I knew that it was rotten.'

Einstein's EPR thought experiment (see page 98) had shown that either there was a big hole in quantum theory or that local reality was untrue. Local reality meant a world that didn't depend on probability and that did not allow distant particles somehow to communicate instantly, Bell came up with his own thought experiment, providing a measurement that would distinguish between the two possibilities. He was a theorist and did not know how this measurement could be put into practice, but he had set a benchmark with what became known as 'Bell's theorem' that made it possible to check the validity of the remarkable claims made by quantum theory. If experimental results fell statistically outside a certain range – known as 'Bell's inequality' – then Bell's theorem was true and local reality was doomed.

Later experiments addressed Bell's theorem and showed that Einstein and, in his heart, Bell were wrong. Quantum theory did appear to be correct and did violate local reality. Bell's untimely and unexpected death at the age of 62 brought to an end the career of a thoughtful and inspired scientist.

Brian Clegg

QUANTUM ENCRYPTION

the 30-second theory

3-SECOND FLASH
Quantum particles, and
particularly entangled
particles, make great
carriers of secret data,
providing their own
random one-time pad.

3-MINUTE THOUGHT
In 2004, Anton Zeilinger,
one of the foremost
experimental physicists
working on quantum
entanglement,
demonstrated the use
of an entangled one-time
pad in a way that beat any
lab experiment for its
dramatic scope. He set up a
600ft (500-metre) link
through the sewers
between City Hall and the
Bank of Austria in Vienna
and (with permission) used
an entanglement-encrypted
message to transfer 3,000
euros from the Mayor of
Vienna's funds to the
University's account.

As long as we have had written
communication, we have tried to keep some
messages secret. Many codes and cyphers
are easily broken, but there is a totally secure
method: the one-time pad. Here a random value
is added to each character to be encrypted.
The result is a truly random piece of text – but
it can be decoded with the key. Such pads are
rarely used because it is too easy for the key to
be discovered by conventional spying. But
quantum physics overcomes this problem.
Quantum cryptography originated with Charles
Bennett and Gilles Brassard, who used the
polarization of individual photons as the key.
This did provide a one-time pad, but had
technical issues that could lay it open to
interception. Quantum entanglement, however,
provides a one-time pad key that does not exist
before the message is sent. Usually the
randomness of the value communicated
instantly by quantum entanglement is a
disadvantage. But if that random value is used
as the key, it will be available to decode the
message as soon as it is encrypted. What's
more, it is possible to test whether particles are
still entangled – so the system automatically
detects any interception of the quantum key.

RELATED THEORIES
See also
QUANTUM SPIN
page 38

EPR
page 98

BELL'S INEQUALITY
page 100

3-SECOND BIOGRAPHIES
CHARLES H. BENNETT
1943–
American physicist and
information theorist mostly
working at IBM

ANTON ZEILINGER
1945–
Austrian quantum physicist
specializing in entanglement
– with a flair for dramatic
demonstrations

GILLES BRASSARD
1955–
French-Canadian computer
scientist and cryptographer

30-SECOND TEXT
Brian Clegg

*Quantum encryption
was used to safe-
guard an electronic
money transfer in
Vienna in 2004.*

QUBITS

the 30-second theory

Electrons have a quantum property called 'spin'. They can spin clockwise or anticlockwise, and it is the electron spin that is responsible for the magnetic properties of certain materials. By hitting electrons with a laser pulse, you can coax them into a superposition state – that is, they assume both quantum states at the same time: rather than having a specific direction of spin they have only probabilities of spinning in each direction simultaneously. The two directions of spin can be assigned values of 0 and 1, corresponding to the 0 or 1 setting of a conventional computer bit, but the superposition and the associated probabilities means that a qubit holds more information. Such a system is called a quantum bit or 'qubit'. Photons, which can be polarized both horizontally and vertically, and atomic nuclei, by assuming two nuclear spin states simultaneously, are other examples of qubits. The superposition state is very delicate: the smallest disturbance, such as an attempt to detect the quantum state of the subatomic particle, causes it to revert back to a non-superposition state, a phenomenon known as decoherence. Quantum entanglement is important in the use of qubits to link data without causing decoherence.

3-SECOND FLASH
Qubits behave like bits. They can be on or off, but in effect they can be on and off at the same time.

3-MINUTE THOUGHT
Qubits will be at the heart of future quantum computers. Any particle or system that can assume two or more quantum states can function as a qubit. Researchers create qubits by several experimental means. For example, they can lock electrons in quantum dots and manipulate their spins with laser beams. The spin of atomic nuclei can be manipulated with radio waves, as in MRI scanners. Serge Haroche pioneered the storing of quantum data by trapped photons in microwave cavities.

RELATED THEORIES
See also
SCHRÖDINGER'S CAT
page 46

DECOHERENCE
page 52

QUANTUM COMPUTING
page 108

MRI SCANNERS
page 124

QUANTUM DOTS
page 130

3-SECOND BIOGRAPHY
SERGE HAROCHE
1944–
French physicist who shared (with David J. Wineland) the 2012 Nobel Prize in Physics for his experimental research in quantum physics

30-SECOND TEXT
Alexander Hellemans

Measuring spin will always produce 'up' or 'down'; which one depends on the qubit's underlying probabilistic state.

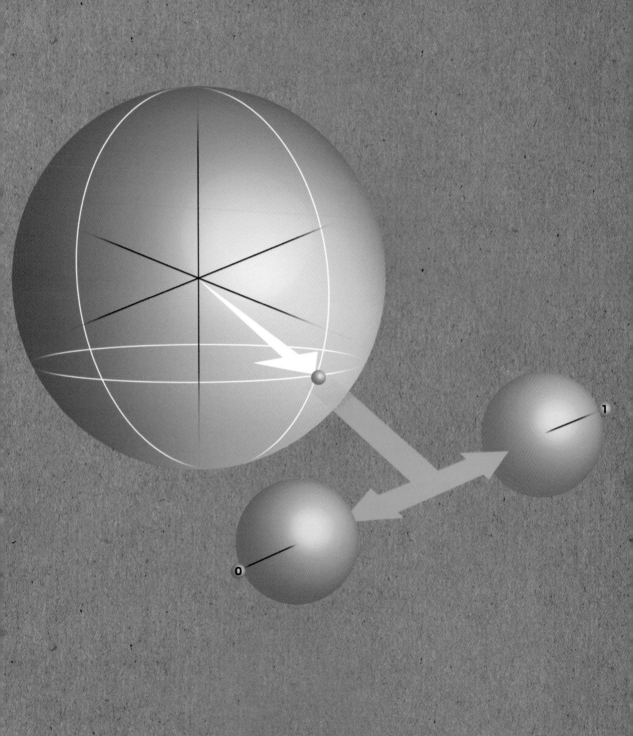

QUANTUM COMPUTING

the 30-second theory

Today's computers contain

millions of tiny transistors that use electric charges to store data as bits. The presence of a charge corresponds to a 1, and its absence to a 0 – this information is called a binary digit, or 'bit'. Computers process numbers by representing them with series of bits that can be switched on or off individually. For example, with four bits you can represent the numbers 0 to 7 as 0000, 0001, 0011, 0111, 1111, 1110, 1100 and 1000. A conventional computer processes these data one at a time. Four cubits, however, with each cubit being a superposition of 0 and 1, will represent these 8 numbers simultaneously, allowing them to be processed in parallel. However, the enormous power of quantum computers becomes apparent when the number of qubits increases. Ten cubits allow the simultaneous processing of 1,023 numbers. The enormous computation power quantum computers are expected to achieve is mindboggling: 20 qubits can process 1 million parallel calculations; with 40 qubits, the number of parallel computations will increase to 1 million million. Although the creation of qubits that remain entangled will require the development of new technologies, researchers are hopeful of using a large number of qubits to achieve enormous computing power.

3-SECOND FLASH
Qubits will play a central role in quantum computers because they will allow parallel processing on a massive scale.

3-MINUTE THOUGHT
Richard Feynman suggested that tiny quantum mechanical computers would be able to simulate quantum systems. Besides the modelling physical process quantum computers will break any records in mathematics. For example, they will be able to factor numbers of 400 digits in a few seconds, enabling the cracking of encryption keys used in banking.

RELATED THEORIES
See also
QUANTUM SPIN
page 38

DECOHERENCE
page 52

BELL'S INEQUALITY
page 100

QUBITS
page 106

QUANTUM DOTS
page 130

3-SECOND BIOGRAPHY
RICHARD FEYNMAN
1918–88
American physicist who suggested computers that would obey quantum mechanical laws

30-SECOND TEXT
Alexander Hellemans

In a quantum computer, quantum bits or qubits replace the traditional bit to enable parallel computation.

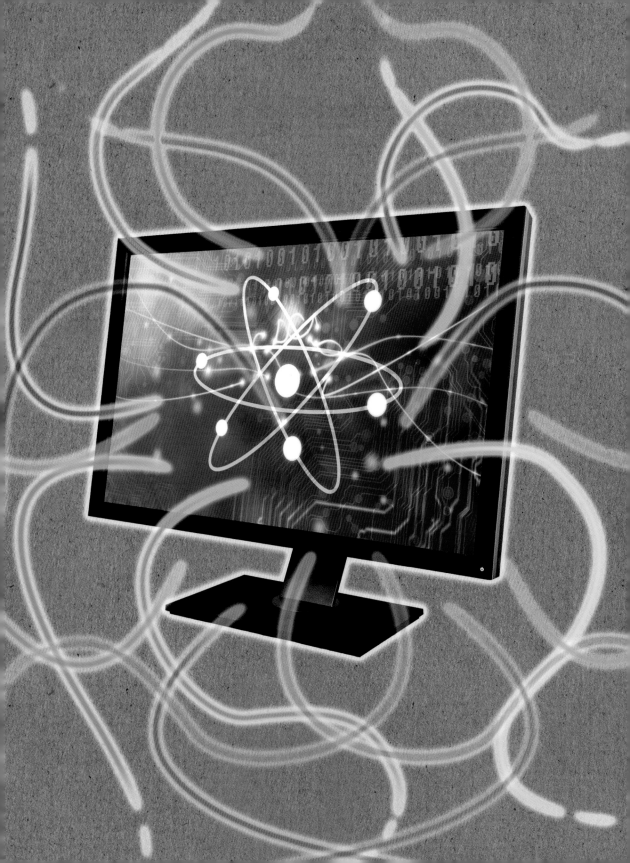

QUANTUM TELEPORTATION

the 30-second theory

On a dark, moonless night in 2012 scientists set the current distance record for quantum teleportation: 89 miles (144 kilometres), using a laser to beam photons between different islands of the Canaries. These photons were intimately connected to one another via the quantum property of entanglement, so that an action made on one of the pair immediately affected its entangled partner, however distant. The team, led by Anton Zeilinger at the University of Vienna, sent one of an entangled pair of photons through the air to a detector on the next island. They then used that pair as a quantum communication line to send information about another quantum object, reconstructing it at the other end of the line. Quantum teleportation sounds much like sci-fi, so when computer scientist Charles Bennett of IBM in New York and colleagues first proposed it in 1993 it attracted immediate attention. It is now a serious area of research, with applications in quantum technologies for computing and telecommunications. It has been demonstrated in various systems, including between clouds of caesium atoms and within electric circuits. Scientists now have their eyes on space: teleporting to orbiting satellites may be essential for a global quantum communications network.

RELATED THEORIES
See also
QUANTUM FIELD THEORY
page 64

BACKWARDS IN TIME
page 72

THE TRANSISTOR
page 120

3-SECOND FLASH
In quantum teleportation, all the information about a quantum object is scanned and recreated in a new place using entangled particles that form the ends of a quantum communication line.

3-MINUTE THOUGHT
Quantum teleportation does not allow faster-than-light communication, because to reconstruct your quantum object at the end of the line you need instructions from the sender, which are sent via a classical communication line. But it does get round the no-cloning rule, which prevents you making a perfect copy of a quantum object. Instead, teleportation works by shifting where the quantum information is located, and in the process destroys the original.

3-SECOND BIOGRAPHIES
CHARLES H. BENNETT
1943–
Fellow at IBM Research whose work focuses on the relation between physics and information

ANTON ZEILINGER
1945–
Austrian quantum physicist and head of the team that pioneered long-distance quantum teleportation experiments

30-SECOND TEXT
Sophie Hebden

Quantum teleportation between the Canary Islands was a precursor to satellite links.

QUANTUM ZENO EFFECT

the 30-second theory

Ancient Greek philosopher Zeno formulated a number of paradoxes that appear to prove that motion is impossible. In terms of classical physics, these paradoxes are easily explained away as fallacies. But in 1977 George Sudarshan and a colleague at the University of Texas drew a parallel between the observation that an arrow in flight does not appear to be moving if we take a single moment in time and a little known quantum phenomenon now termed the quantum Zeno effect. As real-world examples tend to be complicated, it is easier to illustrate the quantum Zeno effect with a thought experiment. The probability that a radioactive atom will decay in a given interval of time is often said to be constant, but this isn't strictly true. Immediately after the atom has been observed in an undecayed state, its rate of decay is zero – although it quickly ramps up to its 'constant' value. But if another observation is made before this has had a chance to happen, the decay rate is pushed back down to zero ... and so on as long as repeated observations are made. It may not be true that a watched kettle never boils – but a watched atom never decays!

RELATED THEORIES
See also
COLLAPSING
WAVEFUNCTIONS
page 50

CONSCIOUSNESS AND
COLLAPSE
page 90

QUANTUM COMPUTING
page 108

QUANTUM BIOLOGY
page 150

3-SECOND BIOGRAPHIES
ZENO OF ELEA
fl. 5th century BCE
Greek philosopher who suggested that precise observation could freeze an arrow in flight

E. C. GEORGE SUDARSHAN
1931–
Indian physicist and prolific researcher in quantum optics and fundamental physics

30-SECOND TEXT
Andrew May

Zeno's arrow did not appear to be moving when watched, and a quantum particle does not decay when observed.

3-SECOND FLASH
If a quantum system is observed frequently enough, it will never change state – even if the system is unstable.

3-MINUTE THOUGHT
Ever since the quantum Zeno effect was discovered, physicists have been trying to build practical applications around it. But it's possible Nature got there first. According to one theory, migratory birds can detect the Earth's magnetic field using pairs of entangled electrons located inside their eyes. What is not clear, however, is how the birds maintain the necessary entangled state sufficiently long for the mechanism to work. The answer may be that they exploit the quantum Zeno effect.

QUANTUM APPLICATIONS

CCD camera A camera based on a charge coupled device, consisting of an array of pixels, each of which builds up an electrical charge as photons hit it.

collimator lens A lens that collects together rays of light to make them aligned.

conduction band The range of energies of an electron in an atom in a material that enables the electron to move freely through the material.

Cooper pairs A pair of fermions (usually electrons) that act as a single particle, bound together by interaction with vibrations in the material that they are passing through. They are responsible for low-temperature superconductivity.

diode laser A laser producing light by stimulated emission from a semiconductor, used in telecommunications, CD and DVD players, laser pointers and printers.

doping Adding an impurity to a semiconductor to change its electrical properties. This makes it much easier for an electron to reach the conduction band or to be accepted into the valence band.

integrated circuit A 'chip' – a thin sheet of semiconductor, usually silicon, with an electronic circuit printed on it.

Josephson junctions A pair of superconductors separated by a thin layer. If a voltage is applied across the junction it causes a high frequency oscillation, providing an extremely accurate measure of the voltage. See page 126.

metamaterial A material specially constructed with unusual electromagnetic properties. Many metamaterials have a negative refractive index, giving them the potential to make unusually powerful lenses or to conceal an object by bending light around it (a process known as 'cloaking').

Moore's law An observation in 1965 by Intel founder Gordon Moore that the capacity of electronic devices roughly doubles annually. Modified to doubling each 18 months or two years, this has proved remarkably accurate, although this now appears to be slowing.

nanoparticles Small pieces of a material around 1–100 nanometres across. Objects at this scale have physical characteristics that are very different from those of larger particles.

Pauli Exclusion Principle The observation that two of the same kind of fermion (electrons, for instance) can't be in the same quantum state at once. For instance, electrons in the same atom can't have the same quantum numbers. See page 58.

photonic lattices A material forming a regular lattice that acts on light as semiconductors do on electrons. They are used to produce high-quality lenses and found in Nature causing the swirls of an opal and the iridescence of a peacock's tail.

photonics Methods of controlling, switching and amplifying light. They are the optical equivalent of electronics.

refractive index A measure of the degree to which a substance bends light as it passes between that substance and another material. It is linked to the velocity of light in the material.

resonance The tendency of a system to vibrate more strongly at particular frequencies. It can be applied to an object like a bell, or to a cavity, like an organ pipe or laser cavity.

semiconductor laser see diode laser

SQUIDs Superconducting Quantum Interference Devices that use Josephson junctions to detect small changes in voltage produced by a shifting magnetic field. They are used to make sensitive magnetometers employed in applications from MRI scanners to unexploded bomb detectors.

stimulated emission The mechanism behind a laser. An atom is pushed into an excited state by a flash of light or electrical current. When hit by an incoming photon, it gives off a second photon with identical frequency. This contrasts with spontaneous emission, which does not involve an incoming photon.

superconductivity The ability of extremely cold materials to conduct electricity without resistance and to expel an electromagnetic field. See page 140.

valence band The band of electrons in an atom still bound to the atom and which is responsible for many of the atom's chemical properties.

THE LASER

the 30-second theory

RELATED THEORIES
See also
PLANCK'S QUANTA
page 18

BALMER'S PREDICTABLE
SPECTRUM
page 22

BOHR'S ATOM
page 24

3-SECOND FLASH
A laser is a light source that works via stimulated or cooperative emission from many atoms to produce a highly organized beam, with all the wavefronts matching up, monochromatic and strongly directional.

3-MINUTE THOUGHT
The laser wielded by Austin Powers' character Dr Evil probably uses carbon dioxide gas as a lasing material, because it emits in the infrared so would fry anything it focuses on. The vast majority of existing lasers are the far less powerful semiconductor or diode lasers that are used for electronic devices and in communications. They typically emit red light and can be built into larger arrays.

We use lasers on a daily basis: they scan our barcodes at the supermarket checkout and they are a key component of CD and DVD players. It may be surprising to learn that these everyday machines are quantum since they depend on the unique energy levels of atoms at their heart. An atom's electrons can be 'excited' to different, precise values by absorbing energy from heat or light. But the atom cannot stay excited all the time, so eventually it releases this energy as a photon of light with a precise frequency and returns to its 'ground' energy state. What happens if an excited atom that is already excited encounters a photon? Instead of absorbing it and releasing it at a random time in a random direction as before, a kind of resonance effect stimulates it to emit a second photon. This photon has exactly the same frequency direction and is in coherence – perfect step – with the incident photon. In a laser a collection of atoms are brought into an excited state by pumping them with a high voltage, so the atoms in an excited state outnumber those in the ground state. Reflecting the emitted photons between mirrors in a cavity stimulates further emissions, generating a powerful laser beam.

3-SECOND BIOGRAPHIES
GORDON GOULD
1920–2005
American physicist who coined the acronym LASER – standing for Light Amplification by Stimulated Emission of Radiation

THEODORE MAIMAN
1927–2007
American physicist who may have invented the first successful optical laser (some credit Gould)

30-SECOND TEXT
Sophie Hebden

Lasers use reflecting cavities repeatedly to stimulate atoms in order to produce coherent photons.

THE TRANSISTOR

the 30-second theory

RELATED THEORIES
See also
THE LASER
page 118

QUANTUM DOTS
page 130

3-SECOND BIOGRAPHIES
WALTER BRATTAIN, JOHN
BARDEEN, WILLIAM
SHOCKLEY
1902–87, 1908–91
& 1910–89
American physicists and
members of the team that
invented the transistor at Bell
Telephone Laboratories in
1947, for which they shared
the 1956 Nobel Prize in Physics

3-SECOND FLASH
Transistors, which make
digital electronics and
computers possible,
exploit the quantization of
electron energy states in
semiconducting materials.

3-MINUTE THOUGHT
Early commercial
transistors, made from
germanium, cost several
dollars each in 1960 and
measured about 12 mm
across. Miniaturization
of transistors made from
silicon has now reached
the point at which about
2 billion can be housed
on a single silicon
microprocessor chip,
at a cost of about 0.0001
cents apiece. This decline
in cost is one aspect of
Moore's law, more usually
expressed in terms of the
number of transistors on
an integrated-circuit chip
– which doubles about
every 18 months.

The peculiarities of quantum
theory, such as uncertainty, superpositions and
superconductivity, are usually found only in
specialized, low-temperature conditions.
However, the consequences of the discrete
quantization of energy states are seen all the
time in, say, the bonds between atoms and the
colours of objects. One of the most significant
technological applications of this quantization
occurs in the transistor, the electronic device
made from a semiconductor that is at the heart
of all digital computing and IT. A semiconducting
material contains electrons in a 'band' of
quantum energy states, rather like a reservoir
entirely filled with water, separated by an energy
gap from another band empty of electrons. If
electrons can gain enough energy to reach the
empty band, they can move around and carry an
electrical current. Only a few electrons can pick
up enough energy from ambient heat to do this
under a transistor's normal operating conditions,
which means that the flow of current can be
finely controlled by doping – adding electrons to
or removing them from the reservoir by
dispersing other kinds of atom into the material
– and by applying electric fields. In this way the
current passing through a transistor can be
controlled and directed electrically, so that it can
act as a switch or an amplifier in digital electronics.

30-SECOND TEXT
Philip Ball

*The use of transistors
has transformed
electronics, from
individual components
to integrated circuits.*

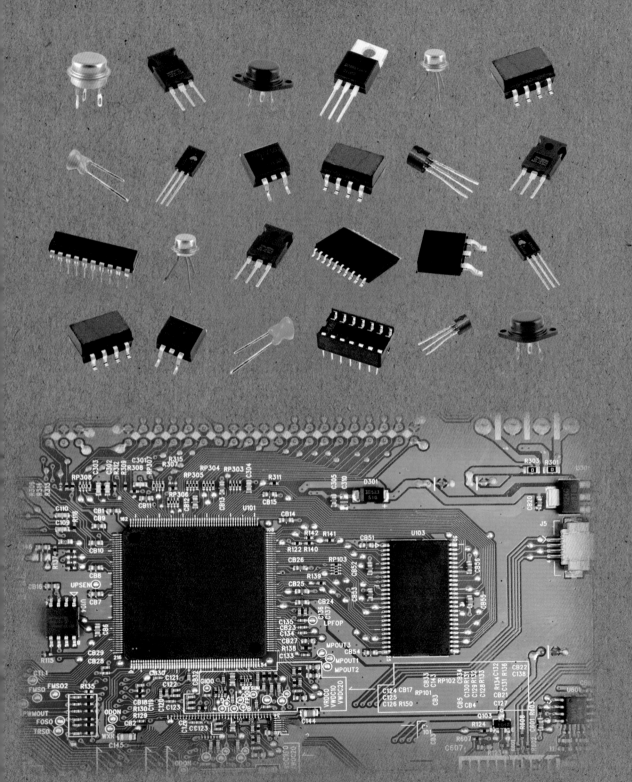

THE ELECTRON MICROSCOPE

the 30-second theory

3-SECOND FLASH
Because electrons have a much shorter wavelength than photons, their amplification is much higher than that of their optical counterparts.

3-MINUTE THOUGHT
The electron microscope demonstrates the dual nature of electrons. When electrons pass through magnetic lenses, their path becomes bent and they behave like particles. Passed through a sample, electrons are diffracted and bend around obstacles, behaving as a wave: they bend to a far lesser extent, resulting in much sharper images. This explains the electron microscope's significance to science: biologists, for instance, can view cell components that were invisible with optical microscopes; nanotechnology would be impossible without it.

Electron microscopes function in a similar way to optical microscopes. In optical microscopes a collimator lens focuses light on a glass slide containing, for example, bacteria. The light is then collected by an objective lens that enlarges the image and focuses it on an eyepiece or CCD camera. In principle, an electron microscope functions in a similar way. But instead of glass lenses, magnets deflect the electrons. A hot cathode produces the electrons, which are then accelerated by an electric field, just like in a cathode ray tube. A magnetic collimator focuses the electrons on the sample, and the electrons that pass through the sample are focused by another set of magnetic lenses on a fluorescent screen, where the electrons form an image that we can see. The resolution of an optical microscope is limited by the wavelength of light, and its magnification is limited to 2,000; anything smaller than a wavelength of light, such as viruses, remains invisible. However, electrons behave as particles, but also as waves. Their wavelength is much smaller than that of light, and therefore electron microscopes can magnify up to 10 million times and 'see' much smaller objects, such as viruses – even atoms.

RELATED THEORIES
See also
WAVE-PARTICLE DUALITY
page 28

3-SECOND BIOGRAPHIES
MAX KNOLL & ERNST RUSKA
1897–1969 & 1906–88
German electrical engineer and physicist who pioneered the apparatus that help construct the electron microscope in 1931

30-SECOND TEXT
Alexander Hellemans

The magnetic field of an electron microscope focuses electrons as a lens does visible light.

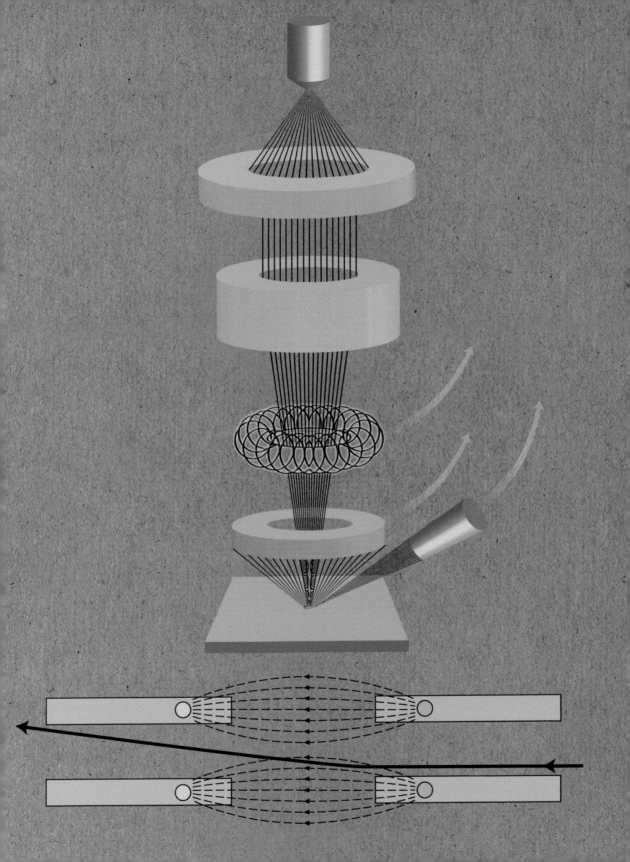

MRI SCANNERS

the 30-second theory

In the 1970s, researchers

including American physician Raymond Damadian, American chemist Paul Lauterbur and British physicist Peter Mansfield developed Magnetic Resonance Imaging (MRI). It provides a non-surgical way of seeing soft tissues inside our bodies that can help diagnose diseases and injuries ranging from cancer to torn ligaments. Inside the scanner, a patient is surrounded by a magnetic field generally 30,000–60,000 times greater than the Earth's magnetic field. This is produced by the scanner's powerful magnet – usually a superconducting electromagnet. Our bodies are 65% water, and every hydrogen atom in this water contains a proton that spins like a top, making each proton behave like a small magnet. The scanner's large magnetic field causes the protons to spin in a particular way. Radio waves directed into the patient's body then alter how the protons spin, and so change their magnetization. Switching these radio waves off again lets the protons 'relax' back to how they were spinning previously, and the signals they emit as they relax are electronically recorded. The time it takes a proton to relax depends on the type of tissue surrounding it. So computer software turns this information and the other detected signals into images revealing the different tissues.

RELATED THEORIES
See also
QUANTUM SPIN
page 38

SUPERCONDUCTORS
page 140

3-SECOND BIOGRAPHIES
HEIKE KAMERLINGH ONNES
1853–1926
Dutch discoverer of superconductivity in 1911

PAUL LAUTERBUR & PETER MANSFIELD
1929–2007 & 1933–
American and British scientists who shared the 2003 Nobel Prize in Physiology or Medicine for their work leading to the development of MRI

30-SECOND TEXT
Sharon Ann Holgate

MRI scanners build a series of cross-sectional images from the electromagnetic radiation given off when radio waves and strong magnetic fields change the spins of protons.

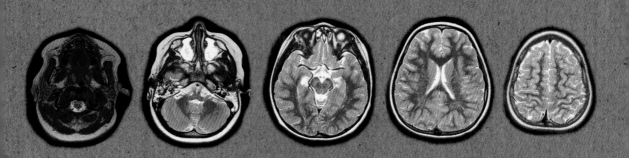

JOSEPHSON JUNCTIONS

the 30-second theory

3-SECOND FLASH
The unexpected appearance of a superconducting current when two superconductors are separated by a tiny gap is promising a host of applications.

3-MINUTE THOUGHT
Because Josephson junctions can function as very fast logic gates, researchers are investigating their application in ultrafast computers. An interesting quantum property of Josephson junctions is that in a superconducting loop they can cause the superposition of two Josephson currents in opposite directions simultaneously. Researchers are currently investigating how tiny superconducting loops can be connected together to store quantum data or to form a quantum computer.

In 1962 Brian Josephson

predicted that Cooper pairs – pairs of linked electrons that travel with no resistance through superconductors – should also tunnel through a non-superconducting or an insulating barrier between two superconductors. Electrons can jump over a tiny gap or insulating layer between two conductors, an effect known as quantum tunnelling. However, unlike electrons tunnelling from one conductor to another conductor at room temperature, Cooper pairs do not require an electric field to coax them through the barrier. All the Cooper pairs in a superconductor share the same wave function, and the difference in the phase of the wave function on each side of the insulating barrier causes the Cooper pairs to tunnel spontaneously through the barrier. However, if a voltage is applied over the junction, the Josephson current is replaced by an oscillating current of a very high frequency. The frequency depends only on the applied voltage. Since frequencies can be measured with higher precision than voltages, Josephson junctions are used as very precise voltmeters. If a Josephson junction is part of a closed loop, the voltage over the junction changes even with extremely weak magnetic fields. Called superconducting quantum interference devices (SQUIDs), they are able to measure the magnetic fields created by the human brain.

RELATED THEORIES
See also
QUANTUM TUNNELLING
page 80

QUBITS
page 106

QUANTUM COMPUTING
page 108

SUPERCONDUCTORS
page 140

3-SECOND BIOGRAPHY
LEO ESAKI & IVAR GIAEVER
1925– & 1929–
Japanese and Norwegian physicists who, along with Brian Josephson, shared the Nobel Prize in Physics in 1973 for their work in electron tunnelling

30-SECOND TEXT
Alexander Hellemans

A superconducting quantum interference device (SQUID) can detect small variations in magnetic fields, including those produced by the brain.

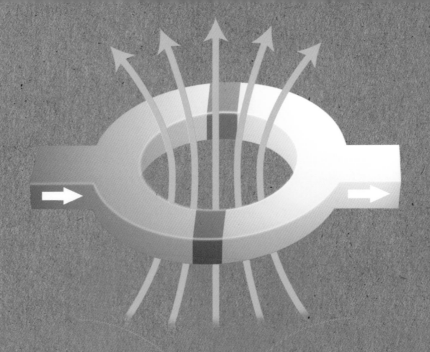

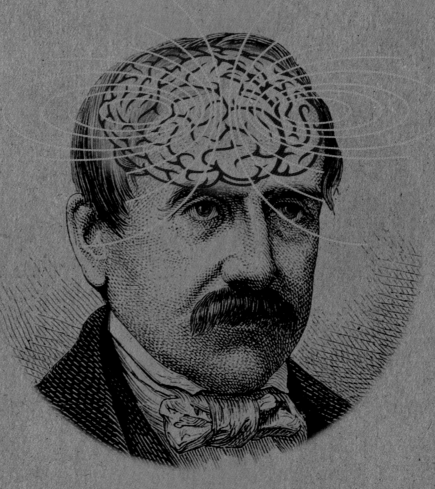

4 January 1940
Born in Cardiff, Wales

1960
Gains Bachelor's degree in Natural Sciences from the University of Cambridge

1962
His paper on what would become known as 'the Josephson Effect' is published in the journal *Physics Letters*

1964
Gains his PhD from Cambridge

1964
The SQUID (superconducting quantum interference device), a highly sensitive magnetometer using Josephson junctions, is invented

1965
Begins a short stint as a research assistant professor at the University of Illinois

1967
Returns to Cambridge as Assistant Director of Research

1972
Becomes Reader (senior lecturer) in Physics at Cambridge

1973
Shares the Nobel Prize in Physics with Leo Esaki and Ivar Giaever

1974
Becomes Professor of Physics at Cambridge

1983
Addresses US Congressional Committee on the subject of 'higher states of consciousness'

1988
Sets up the Mind–Matter Unification Project at Cambridge

2007
Retires from his Cambridge professorship, but continues in active research

BRIAN JOSEPHSON

Like many great physicists,
Brian Josephson showed an intuitive grasp
of the subject from an early age. As an
undergraduate at Cambridge, he constantly
impressed his teachers with the depth of his
understanding, and by the time he obtained his
first degree – at the unusually young age of 20
– he had already published his first research
papers. Choosing to remain at Cambridge,
he pursued a PhD in the area of physics that
intrigued him the most: a phenomenon called
superconductivity that occurs at extremely
low temperatures. He was still working on his
PhD when he wrote the paper for which he
is most famous, 'Possible New Effects in
Superconductive Tunnelling'. It was known that
quantum theory allows particles to 'tunnel'
through otherwise impenetrable barriers, but
Josephson showed how this could produce a
hitherto unknown effect – since dubbed the
Josephson effect – in the context of
superconductors. This was one of the first
examples of a quantum effect that can be
exploited at macroscopic scales, and it led to
the now well-established technology of the
Josephson junction.

Josephson was still only 33 when he shared
the 1973 Nobel Prize in Physics with two other
researchers who had worked on quantum
tunnelling. Although he was the youngest of
the three, he received half the prize while the
other two received a quarter each. It was only
in the following year, 1974, that Josephson was
finally made a full professor at Cambridge – a
post he held until his retirement in 2007.

By the late 1970s Josephson was becoming
disillusioned with mainstream physics, feeling
that it ignored large areas of experience which,
given the chance, might be illuminated by
quantum theory, or a yet-to-be-discovered
extension of it. He developed an interest in
Eastern philosophy, meditation and higher
states of consciousness, and in 1988 – still
under the auspices of the physics department
at Cambridge – he set up his long-running
Mind-Matter Unification Project. This was
concerned with such subjects as language,
music and cognition – which, although
respectable topics in other academic
departments, were not normally the province
of a theoretical physicist. In more recent years
Josephson has written on distinctly non-
academic subjects, including telepathy and
homeopathy, which has inevitably brought him
into conflict with his more conservative peers.
Josephson himself remains open-minded about
what he calls 'heretical science'. On his website
he adopts the motto of the Royal Society,
nullius in verba, which he paraphrases as 'take
nobody's word for it'.

Andrew May

QUANTUM DOTS

the 30-second theory

3-SECOND FLASH
Quantum dots are nanoparticles, often attached to a substrate or active surface, which, because of their small size acquire quantum properties that are of interest to technology.

3-MINUTE THOUGHT
The ability of quantum dots to eject two electrons simultaneously when hit by a single photon, means they offer the possibility of enhancing the efficiency of solar cells. Because they can be tuned to emit any colour of the spectrum, researchers are investigating how to use them in displays and in light-emitting diodes.

Among the factors limiting the downsizing of structures on electronic chips are quantum effects that interfere with the functioning of the circuitry. For example, when conductors come too close, electrons tunnel from one conductor to the other. However, researchers are trying to turn these quantum effects to their advantage. Quantum dots are tiny nanoparticles made of semiconducting materials including silicon, cadmium selenide, cadmium sulphide and indium arsenide. They are designed to be so small that quantum effects become apparent. Typically the size of 10–50 atoms (2–10 nanometres), they start behaving like atoms themselves. The electrons in the conduction bands start populating discrete quantum levels imposed by the Pauli Exclusion Principle. Therefore quantum dots are sometimes called 'artificial atoms'. Because nanodots consist of semiconducting materials, there is a gap between the conduction band and the highest band. Photons can excite electrons in the valence band and bump them into the conduction band. These electrons can then jump back to the valence band while emitting a photon. Now the energy difference between the conduction band and the valence band can be tuned by changing the size of the nanoparticles, where the energy difference is the highest for the smallest particles.

RELATED THEORIES
See also
THE PAULI EXCLUSION PRINCIPLE
page 58

QUANTUM COMPUTING
page 108

THE TRANSISTOR
page 120

3-SECOND BIOGRAPHY
WOLFGANG PAULI
1900–58
Austrian theoretical physicist who introduced the exclusion principle that took his name

30-SECOND TEXT
Alexander Hellemans

Quantum dots could be used to produce visual displays with an unequalled level of fine detail.

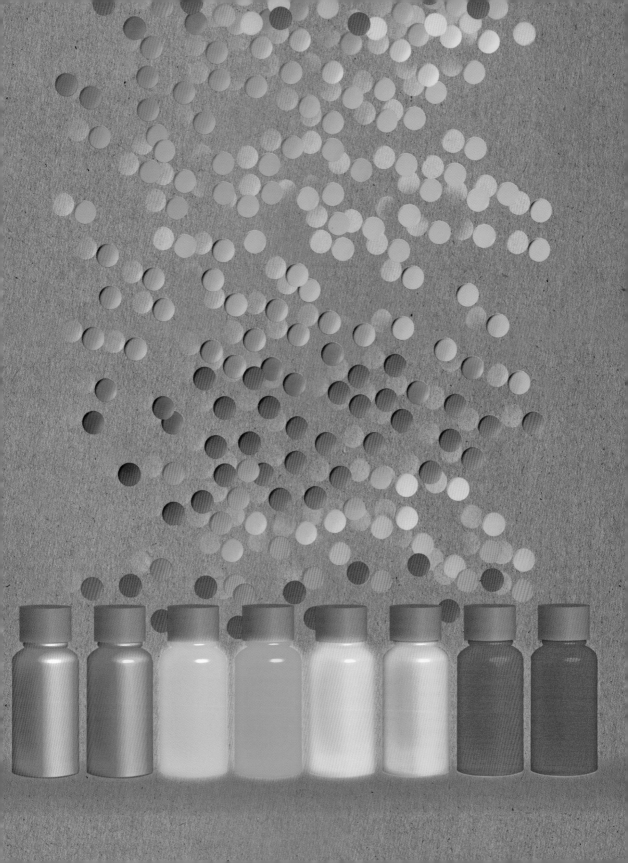

QUANTUM OPTICS
the 30-second theory

All optical devices work at the quantum level, and the interaction between photons of light and the atoms of objects, mirrors and lenses is explained by QED (quantum electrodynamics). However, recently there are many more direct applications of quantum theory in optics, sometimes known as photonics. Among the most dramatic possibilities here are quantum lenses – materials that manipulate photons in ways different from traditional lenses. Take, for instance, metamaterials. These substances have complex structures – for example, layers of lattices or patterns of tiny holes in a metallic sheet – that produce strange effects like a negative refractive index, bending light the opposite way to conventional lenses or prisms. Their structures give metamaterials the ability to focus on far smaller objects than a conventional lens, producing so-called super-lenses. Another example of a quantum optical structure is a photonic lattice, which acts on light in a way similar to a semiconductor acting on electrons. The iridescence of some butterfly wings is produced by natural photonic lattices. Photonic lattices could be employed in future optical computers, and crystals of these lattices are already used in special paint systems, reflection-reducing coatings on lenses and in high transmission photonic fibre optics.

RELATED THEORIES
See also
QED BASICS
page 66

FEYNMAN DIAGRAMS
page 70

BEAM SPLITTERS
page 78

3-SECOND BIOGRAPHIES
VICTOR VESELAGO
1929–
Russian physicist and the first to consider negative refractive index and metamaterials

JOHN PENDRY
1943–
British theoretical physicist who worked on theories of metamaterials for perfect lenses and invisibility cloaks

ULF LEONHARDT
1965–
German physicist working on practical invisibility cloaks

30-SECOND TEXT
Brian Clegg

Photonic lattices, as in iridescent butterfly wings, manipulate light through their complex structure.

3-SECOND FLASH
All optics involve quantum phenomena, but special quantum optical materials like metamaterials and photonic lattices manipulate photons in a way that conventional optics cannot duplicate.

3-MINUTE THOUGHT
Because of their negative refractive index metamaterials can bend light around an object, making it disappear. This has already been achieved on a small scale, but is limited, as the materials used absorb too much light to provide total invisibility. However, there are alternative mechanisms that either optically amplify the restricted output of the metamaterial, or use a photonic crystal to control the way the light is diffracted, so we may have Harry Potter-style cloaking in the not-too-distant future.

QUANTUM EXTREMES

absolute zero The lowest possible temperature at which the atoms in a material are all in their lowest energy state. This cannot be reached in practice in a finite number of steps. Its value is −273.15°C (0 K).

antiquarks Antimatter equivalents of the fundamental quark particles that make up neutrons and protons and hence the bulk of matter. Each quark has an antiquark that is identical to it, apart from having the opposite charge and 'colour' (this being an additional property of quarks).

asymptotic freedom The unusual behaviour of the strong force between quarks, which becomes weaker as the particles get closer and stronger as they separate, meaning that individual quarks are never seen.

black hole A location at which mass has been made so compact that it collapses to a point under gravitational pull, where not even light can escape. Black holes are most frequently formed by the collapse of a massive star. The apparent size of the black hole is its 'event horizon', which is the distance from the centre where nothing can escape, though the black hole itself is a singularity, a dimensionless point.

boson A particle that obeys Bose-Einstein statistics (as opposed to a fermion). Typically bosons are the particles that carry forces – most notably photons and the famous Higgs boson, but the term also applies to atomic nuclei with an even number of particles. Unlike fermions, many bosons can be in the same state simultaneously.

Casimir effect A quantum effect causing an attraction between two parallel plates that are very close together, which can be explained by the limitations caused by the very narrow gap on the formation of virtual particles, compared with outside the plates (putting pressure on them from outside) or in terms of limitations on the zero point energy – the energy of empty space.

CERN The European Organization for Nuclear Research, based near Geneva on the Swiss/French border. CERN is home to the Large Hadron Collider (LHC) and to a wide range of other particle experiments, including significant antimatter research.

dark state A state in which an atom cannot absorb or emit a photon, which can result in some kinds of Bose-Einstein condensate (see page 144), 'trapping' light.

glueballs A hypothetical particle made up solely of gluons.

gluons The elementary boson particles that carry the strong force between quarks, much as photons carry the electromagnetic force between charged particles. Unlike photons, which have no electrical charge, gluons come in different 'colours', the additional charge type associated with quarks.

graviton The hypothetical particle that would carry the gravitational force in a quantum theory of gravity, just as the photon carries the electromagnetic force.

lambda point The temperature below which helium goes from being a normal fluid to a superfluid.

Meissner effect The expulsion of any magnetic field from a superconductor (one of its defining properties), which means that a magnet will levitate above a superconductor.

quarks The fundamental particles making up neutrons and protons, hence the bulk of all matter. Quarks can come in six different flavours: up, down, strange, charm, bottom and top (these names have no significance). Neutrons and protons are different combinations of three up and down quarks; other particles are formed from pairs of quarks.

singularity A point in spacetime at which a property becomes infinite (most commonly applied to a black hole, where the gravitational field becomes infinite) and current theories break down. The heart of a black hole is a singularity.

spacetime Relativity treats time as a fourth dimension. In relativity there is no absolute position or absolute time because the way things move influences their position in time, so it is necessary to consider spacetime as a whole, rather than to think of space and time independently.

specific heat capacity The amount of heat energy required to make a fixed change in temperature in a given mass of a substance.

superconductivity The ability of extremely cold materials to conduct electricity without resistance and to expel an electromagnetic field. See page 140.

ZERO POINT ENERGY
the 30-second theory

The energy-time formulation of Heisenberg's Uncertainty Principle allows large energy fluctuations to exist for brief periods of time. One consequence of this is that every quantum system has a minimum energy state – known as its 'zero point energy' – below which it is impossible to get. Even the quantum fields permeating the vacuum of space have a zero point energy. So empty space is not really empty! It can be pictured as an ever-changing sea of 'virtual particles' flickering in and out of existence. No one really knows how much energy there is in the vacuum: some theoretical predictions suggest there ought to be a huge amount, while observations of the large-scale structure of the universe imply a much smaller value. Despite uncertainties over its magnitude, there is tantalizing experimental evidence that the zero point energy of the vacuum does indeed exist. An oft-cited example is the Casimir effect, which gives rise to an attractive force between two closely spaced metal plates. This phenomenon, predicted by Hendrik Casimir in 1948, has since been demonstrated in the laboratory. But does the Casimir effect prove the existence of zero point energy? Casimir himself thought it did, but there are alternative explanations – so the jury is still out.

3-SECOND FLASH
Thanks to Heisenberg's Uncertainty Principle, there is no such thing as empty space – even the vacuum is a seething mass of virtual particles.

3-MINUTE THOUGHT
Can the world's energy problems be solved by extracting zero point energy from the vacuum? Most physicists would say this is impossible – how can you extract energy from what is, by definition, the minimum energy state? Nevertheless, in 1984 Robert L. Forward proposed a thought experiment based on the Casimir effect that showed how a 'vacuum fluctuation battery' might work … although in practice the system would probably end up using more energy than it generated.

RELATED THEORIES
See also
HEISENBERG'S UNCERTAINTY PRINCIPLE
page 48

QUANTUM FIELD THEORY
page 64

3-SECOND BIOGRAPHIES
HENDRIK CASIMIR
1909–2000
Dutch physicist who worked for Niels Bohr and Wolfgang Pauli

ROBERT L. FORWARD
1932–2002
American physicist, aerospace engineer and science-fiction writer

30-SECOND TEXT
Andrew May

In the Casimir effect, two narrowly separated metal plates in a vacuum are forced towards each other.

SUPERCONDUCTORS

the 30-second theory

In 1911 Dutch physicist Heike
Kamerlingh Onnes was testing the properties of mercury at low temperatures. To his amazement, at 4.2 K (−268.95°C) its electrical resistance simply disappeared. If a current was started in such a 'superconductor', it would flow as long as the material was kept sufficiently cold. This was explained in the 1950s by American physicists John Bardeen, Leon Cooper and John Schrieffer. They discovered that at low temperatures electrons form Cooper pairs – twin electrons that bind together to form a single boson particle. Such particles, like photons, can collect together in the same state, whereas a fermion like a normal electron can only have a single particle in the same location and state. At very low temperatures a 'condensate' forms – a collection of Cooper pairs acting as a single entity. Instead of individual electrons flowing through the conductor there is a single end-to-end substance, not resisted by the atoms of the material. Superconductivity makes possible the hugely powerful magnets in MRI scanners and particle accelerators like the Large Hadron Collider. It also produces the Meissner effect, where a magnet 'levitates' above a superconductor because the superconductor prevents the magnetic field extending inside it.

RELATED THEORIES
See also
JOSEPHSON JUNCTIONS
page 126

SUPERFLUIDS
page 142

BOSE-EINSTEIN CONDENSATES
page 144

3-SECOND BIOGRAPHIES
HEIKE KAMERLINGH ONNES
1853–1926
Dutch physicist, first to liquefy helium and to observe superconductivity. Won the 1913 Nobel Prize in Physics for investigating matter at low temperatures

JOHN BARDEEN, LEON N. COOPER & JOHN R. SCHRIEFFER
1908–91, 1930– & 1931–
American physicists who shared the 1972 Nobel Prize in physics for superconductivity theory

30-SECOND TEXT
Brian Clegg

Superconducting magnets produce levitation and are used to produce powerful magnetic fields for accelerators.

3-SECOND FLASH
A change in the quantum behaviour of electrons at very low temperatures produces a current that flows with no resistance, used to produce extremely powerful electromagnets.

3-MINUTE THOUGHT
The Cooper-pair mechanism works only at very low temperatures, and it was thought that this would make it impossible for a superconductor to operate above 30 K (−243.15°C). However since the 1980s 'high-temperature' superconductors operating at up to 135 K (−138.15°C) have been produced and the hope is that a room-temperature superconductor could be made. It remains uncertain how high temperature superconductors work, though the effect is thought to relate to electron spin.

SUPERFLUIDS

the 30-second theory

Heike Kamerlingh Onnes, who discovered superconduction, was also the first physicist to liquefy helium. In 1911 he established that below 2.17 Kelvin (−270.98°C) the conduction of heat by liquid helium increased dramatically. This temperature became known as the lambda point because of the shape of the graph of heat conductivity in function of the temperature. Why this happened remained a mystery until 1938 when Pyotr Kapitsa in Russia and John Allen and Don Misener in the UK discovered that at a temperature below the lambda point the viscosity of helium disappeared completely. The explanation of the phenomenon has a lot in common with the cause of superconduction in metals. In some superconductors, electrons, which are bosons, pair up and loose any electric resistance. Helium-4 atoms form a Bose-Einstein Condensate (BEC) whereby all the atoms share the same wave function, and lose any mechanical friction. Surprisingly, helium-3 – which consists of fermions and, obeying the Pauli Exclusion Principle, cannot form a BEC – can also become superfluid when cooled to 0.002 above absolute zero (−273.15°C). Here a mechanism closer to the one that causes superconduction is responsible. The helium-3 atoms pair up by aligning their magnetization, forming pairs that are bosons.

3-SECOND FLASH
Superfluidity is the only quantum effect that can be observed by the naked eye: a stirred superfluid will keep rotating forever.

3-MINUTE THOUGHT
Superfluid helium is a real 'Houdini'. Because of the absolute absence of friction, it will flow over the rim of a glass containing it, and it will pass through the smallest hole, just a few atoms wide. Superfluid helium was used for the first time for cooling a telescope mirror in space with the Infrared Astronomy Satellite, operational in 1983. The diffusion of the helium through two brass plugs kept the temperature at 1.6 Kelvin (-271.55°C).

RELATED THEORIES
See also
SUPERCONDUCTORS
page 140

BOSE-EINSTEIN
CONDENSATES
page 144

3-SECOND BIOGRAPHIES
HEIKE KAMERLINGH ONNES
1853 –1926
Dutch physicist and pioneer of refrigeration techniques who investigated the behaviour of matter at very low temperatures

PYOTR KAPITSA
1894–1984
Russian physicist, awarded the 1978 Nobel Prize in physics for discovery of superfluidity; parallel work by Canadian physicists John Allen and Donald Misener did not result in a share of the prize

30-SECOND TEXT
Alexander Hellemans

A superfluid has no viscosity and once rotating will continue doing so indefinitely.

BOSE-EINSTEIN CONDENSATES

the 30-second theory

We are taught that there are three states of matter – solid, liquid and gas – and may be familiar with a fourth in a plasma, like a gas but composed of charged ions. But physics recognizes a fifth, purely quantum state, a Bose-Einstein condensate. This only forms when a collection of appropriate atoms is cooled close to absolute zero (−273.15°C). Atoms can either be fermions, which (like electrons and protons) have to be in unique states to exist in close proximity, or bosons, which (like photons of light) can have many identical particles crammed together. A Bose-Einstein condensate consists of bosons, mostly in their lowest energy state due to the extreme cooling, which begin to behave collectively like a large-scale version of a quantum particle. This produces effects like superfluidity, and also means that large collections of atoms can undergo the quantum processes usually only associated with individual particles, like the quantum mechanical interference produced by the double-slit experiment. As yet there are no applications of Bose-Einstein condensates, though it has been suggested they could be used to detect stealth aircraft by monitoring how small changes in gravity influence quantum interference.

The wavefunctions of interacting condensates produc interference patter which would be influenced by chang in gravity, and able to detect a passing stealth aircraft.

1 January 1894
Born in Calcutta, son of Surendra Nath and his wife Amodini

1913
Graduates from Calcutta University

1917
Begins lecturing at Calcutta University

1919
Translates into English Albert Einstein's papers on relativity

1921
Becomes Reader in Physics at Dhaka University

1924
Publishes paper on the indistinguishability of identical quantum particles

1925
Travels to Europe, where he works with Marie Curie and meets Einstein

1926
Becomes Professor and Head of Physics at Dhaka University

1945
Becomes Professor of Physics at Calcutta University, and President of the Indian Physical Society

1948
Launches the Science Association of Bengal

1954
Enters upper house o the Indian parliament Indian Government honours him with its second highest civilia award, the title Padm Vibhushan

1958
Elected a Fellow of th Royal Society

4 February 1974
Dies in Calcutta

SATYENDRA NATH BOSE

An award-winning mathematics and physics student, Satyendra Nath Bose was a polymath who also excelled in languages. In 1919, while lecturing in physics at Calcutta University, Bose and fellow physicist Meghnad Saha co-produced the first English translations of Einstein's groundbreaking papers on relativity. Five years later the 30-year-old Bose, by then at Dhaka University, made his own breakthrough.

In 1900, Max Planck proposed that the heat given out by very hot objects comes in separate chunks or 'quanta' of energy. Although it explained the distribution of energy through the spectrum of this thermal radiation, some of the mathematical arguments Planck used to derive his theory – known as Planck's Law – were unconvincing. Bose solved these problems by deriving Planck's Law in a different way. Whereas Planck thought the individual quanta could be distinguished from one another, Bose assumed you couldn't tell them apart. Bose's breakthrough idea – the indistinguishability of identical quantum particles – became one of the cornerstones of quantum theory.

Bose sent his 1924 paper on the subject, entitled 'Planck's Law and the Light Quantum Hypothesis', to Albert Einstein after a British science journal had rejected it. Impressed, Einstein endorsed the paper and translated it into German. It was duly published in the physics journal *Zeitschrift für Physik*. Einstein subsequently added to Bose's ideas, and their combined work became known as Bose–Einstein statistics. Particles obeying these statistics are called bosons in recognition of Bose's groundbreaking contribution.

Einstein also helped Bose to obtain a visa to go to Europe. Bose worked with Marie Curie in Paris, and met Einstein in Berlin, along with other founders of quantum theory including Bohr, Heisenberg and Schrödinger.

Despite his formidable intellect, Bose could appear disorganized in both his scientific research and personal appearance. He would jump from one area of study to another, and one photograph reveals him wearing a beret and neckerchief with traditional Indian clothes. Bose, who also studied poetry and literature, and played the *esraj* (Indian stringed instrument), was renowned for his friendliness, modesty and compassion. To help India compete on an international platform, he campaigned to have science taught in native languages instead of English and launched the Science Association of Bengal, which published a monthly Bengali science magazine.

Bose helped to change the course of 20th-century physics, and his influence is still being felt long after his death. In 1995, the first Bose-Einstein condensate – a large number of atoms governed by Bose–Einstein statistics, which are cooled to just above absolute zero and behave like one giant atom – was produced.

Sharon Ann Holgate

QUANTUM CHROMODYNAMICS

the 30-second theory

3-SECOND FLASH
Colour charge, a property similar to electric charge, is carried by quarks and gluons and is the source of a strong force that builds the nuclei of atoms.

3-MINUTE THOUGHT
QCD explains why nuclear forces between neutrons and protons are very strong, whereas those between quarks and gluons, at very short distances, are relatively feeble. It enables physicists at CERN to analyse collisions between protons in terms of their quark and gluon constituents and aided the discovery of the Higgs boson. QCD implies that glueballs exist, but concealed among particles made of quarks and antiquarks, they are hard to isolate.

The strong nuclear forces acting on protons and neutrons defied theoretical description until around 1970, when a deeper layer of reality was discovered: protons and neutrons are made of quarks. The quarks carry electric charge and also another form of charge, whimsically named 'colour'. This colour is the source of the forces that grip quarks in clusters, forming first neutrons and protons and then atomic nuclei. The relativistic quantum theory of the forces between particles that carry or contain colour charge is known as quantum chromodynamics – QCD. The theory is similar to quantum electrodynamics, QED, which describes the interactions between electrically charged particles and light. Light is itself described as made of particles – photons. The analogous cases in QCD involve the interactions between particles carrying colour charge, and 'gluons' – the analogues of photons. However, whereas photons carry no electric charge and travel independently of one another, gluons carry colour charge and interact with one another in flight. This causes the forces arising in QCD to behave quite differently from the electromagnetic force of QED. For example, quarks are forever confined within particles such as protons and cannot exist in isolation – unlike electrons, which do not carry colour charge and can escape from atoms.

RELATED THEORIES
See also
QUANTUM FIELD THEORY
page 64

QED BASICS
page 66

3-SECOND BIOGRAPHIES
DAVID GROSS, H. DAVID POLITZER & FRANK WILZCEK
1941–, 1949– & 1951–
American physicists and joint recipients, of the 2004 Nobel Prize in Physics for their discovery that QCD is a viable theory of the interactions between quarks

30-SECOND TEXT
Frank Close

The colours of quantum chromodynamics are not visible colours but a property similar to an electric charge.

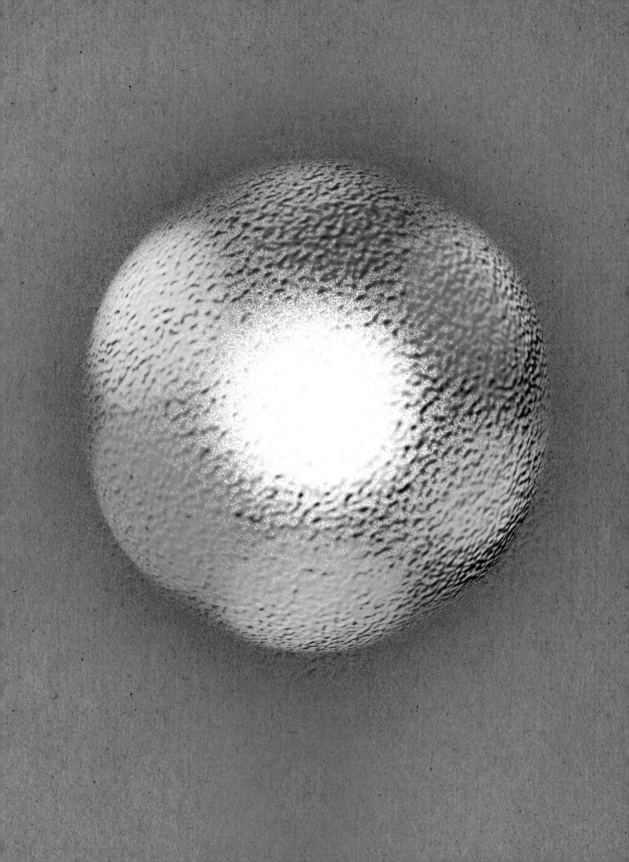

QUANTUM BIOLOGY

the 30-second theory

3-SECOND FLASH
Quantum-mechanical effects such as tunnelling, superposition and entanglement are found to play a role in some biological processes, such as photosynthesis and navigation by birds.

3-MINUTE THOUGHT
A speculative and controversial role for quantum effects in biology was proposed in the 1990s by physicist Roger Penrose and physician Stuart Hameroff. They suggested that human consciousness arises from superpositions of the quantum states of protein fibres in neurons, called microtubules. Wavefunction collapse in these states, they posit, would enable the brain to perform a kind of quantum computation that discloses answers to questions not provable by the formal rules of logic.

Could quantum effects play a role in biology? At face value it seems unlikely: quantum effects are typically fragile and appear only in well-isolated conditions and at low temperatures, whereas life is warm, wet and messy. But quantum behaviour does occur in biological systems, and these are explored in a new discipline: quantum biology. Many biochemical reactions controlled by enzymes involve the movement of an ionized hydrogen atom – a lone proton – from one molecule to another. The proton is so light, it can move by quantum tunnelling through an energy barrier rather than having to climb over it. It remains unclear if such tunnelling in enzyme chemistry is incidental or has been exploited by natural selection. More strikingly, when the light-harvesting molecules involved in bacterial photosynthesis capture sunlight, the energy seems to be distributed among a superposition of quantum waves that move in step (coherently) towards the place where the initial photosynthetic chemical reaction occurs, enhancing the efficiency of this energy transport. And the biochemical magnetic compass that enables some birds to navigate using the Earth's magnetic field may involve quantum entanglement between electrons. Quantum tunnelling has also been proposed as having a role in the molecular mechanism of smell.

RELATED THEORIES
See also
QUANTUM TUNNELLING
page 80

3-SECOND BIOGRAPHIES
GRAHAM FLEMING
1949–
American chemist who first observed quantum coherence in photosynthetic energy transfer in 2007

PAUL DAVIES
1946–
English physicist whose interest in the physics of life ranges from astrobiology to the mechanisms of cancer and quantum effects in biology

30-SECOND TEXT
Philip Ball

Quantum effects could be responsible for the ability of some birds to navigate using magnetic fields.

QUANTUM GRAVITY

the 30-second theory

The quest for a theory of quantum gravity is one of physics' biggest challenges today. The hope is to unite the two pillars of modern physics: Einstein's space-warping theory of gravity (general relativity), and quantum theory, which describes the atomic realm, into a grand 'theory of everything'. A successful theory would help explain the first moments of the big bang, and what happens near the singularity at the centre of a black hole. Creating such a theory is not easy. Whereas the maths of quantum theory treats time and space as an impassive, unchanging background against which events take place, in relativity, they are ... relative. Worse, to test any theory of quantum gravity you need to look at extremely high energies and long distances, currently beyond our experimental reach. Nevertheless, some physicists are exploring different approaches. The most popular idea is string theory, which describes elementary particles as tiny, vibrating loops of energy that inhabit a space of 9 or 10 dimensions. Another possibility is that space has a discreet microstructure of discrete edges joined by nodes, called a spin network. These nodes can entwine to form twisted loops that comprise elementary particles in the theory of Loop Quantum Gravity.

RELATED THEORIES
See also
QUANTUM FIELD THEORY
page 64

3-SECOND FLASH
Theories of quantum gravity attempt to merge general relativity – which describes gravity as a geometric property of space and time – with the quantum physics of the atomic realm.

3-MINUTE THOUGHT
Many ideas about quantum gravity point to discrete structures in spacetime, whose cumulative effect on particles travelling vast distances through the universe may be observable. If gravity can indeed be quantized, then finding the quantum unit of gravity – the graviton – would be evidence enough. But gravity is the weakest of the fundamental forces of nature, so detecting a graviton is extremely hard; some say it would require a detector more massive than a black hole.

3-SECOND BIOGRAPHIES
PAUL DIRAC
1902–84
British physicist and the first person to attempt to quantize Einstein's theory of general relativity in 1932

ABHAY ASHTEKAR
1949–
Indian physicist who made an important advance in 1986 by reformulating general relativity to bring its mathematical language closer to that used in particle and quantum physics

30-SECOND TEXT
Sophie Hebden

The big bang and black holes cannot be properly understood without a quantum theory of gravity.

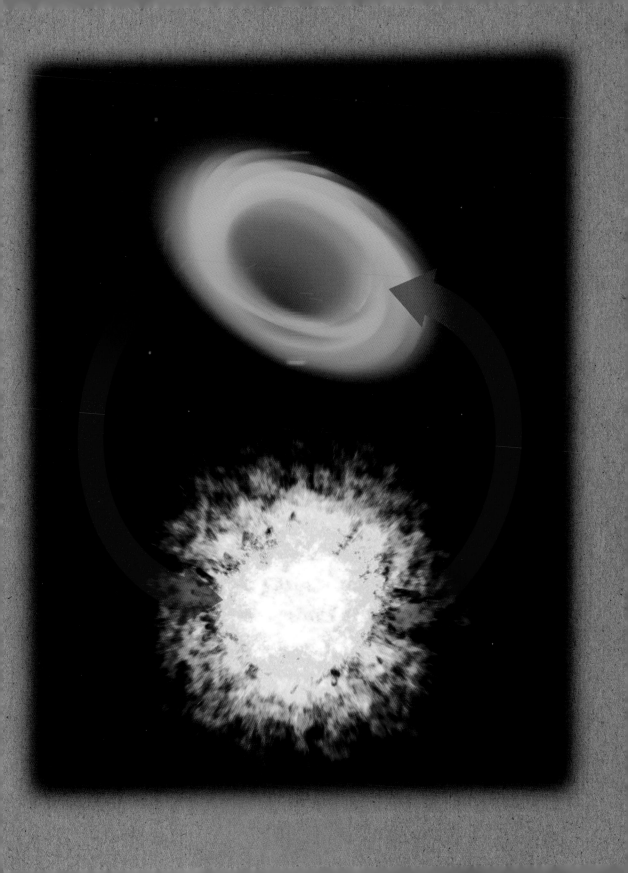

RESOURCES

BOOKS

Paradox: The Nine Greatest Enigmas in Physics
Jim Al-Khalili
(Black Swan, 2013)

The Many Worlds of Hugh Everett III
Peter Byrne
(Oxford University Press, 2010)

Quantum Theory Cannot Hurt You
Marcus Chown
(Faber & Faber, 2008)

The God Effect: Quantum Entanglement
Brian Clegg
(St Martin's Griffin, 2009)

The Quantum Age
Brian Clegg
(Icon Books, 2014)

Antimatter
Frank Close
(Oxford University Press, 2010)

The Infinity Puzzle: Renormalisation and Quantum Theory
Frank Close
(Oxford University Press, 2011)

Nothing
Frank Close
(Oxford University Press, 2009)

The Quantum Universe
Brian Cox & Jeff Forshaw
(Allen Lane, 2011)

QED: The Strange Theory of Light and Matter
Richard Feynman
(Penguin, 1990)

Computing with Quantum Cats
John Gribbin
(Bantam, 2013)

Erwin Schödinger and the Quantum Revolution
John Gribbin
(Black Swan, 2013)

Beam: The Race to Make the Laser
Jeff Hecht
(Oxford University Press, 2010)

The Amazing Story of Quantum Mechanics
James Kakalios
(Duckworth, 2010)

Quantum
Manjit Kumar
(Icon Books, 2009)

ARTICLES & WEB SITES

Biography of Sir Peter Mansfield
www.nobelprize.org/nobel_prizes/
medicine/laureates/2003/mansfield-bio.
html

The Higgs boson: One year on
By CERN particle physicist Pauline Gagnon
home.web.cern.ch/about/updates/2013/07/
higgs-boson-one-year

Information for the public on the 2001
Nobel Prize in Physics for Bose-Einstein
condensation from the official website of
the Nobel Prize
www.nobelprize.org/nobel_prizes/physics/
laureates/2001/popular.html

Institute of Physics: Quantum Physics
www.quantumphysics.iop.org

Jim Al-Khalali - Quantum life (video)
www.richannel.org/jim-al-khalili--
quantum-life-how-physics-can-
revolutionise-biology

New Scientist 'Quantum World'
topic guide
www.newscientist.com/topic/quantum-
world

Quantum physics news
www.sciencedaily.com/news/matter_
energy/quantum_physics/

Robert Peston learns quantum physics:
www.brianclegg.blogspot.co.uk/2013/08/
peston-physics.html

Royal Society – why is quantum physics
important?
invigorate.royalsociety.org/ks5/the-best-
things-come-in-small-packages/why-is-
quantum-physics-important.aspx

Scientific American
www.scientificamerican.com/topic/
quantum-physics/

S N Bose Project by
Bose's grandson Falguni Sarkar
www.snbose.org

NOTES ON CONTRIBUTORS

Philip Ball is a freelance writer, and was an editor for *Nature* for more than 20 years. Trained as a chemist at the University of Oxford, and as a physicist at the University of Bristol, he writes regularly in the scientific and popular media, and has authored books including *H2O: A Biography of Water*, *Bright Earth: Art and the Invention of Colour*, *The Music Instinct: How Music Works and Why We Can't Do Without It* and *Curiosity: How Science Became Interested in Everything*. His book *Critical Mass: How One Thing Leads to Another* won the 2005 Aventis Prize for Science Books. He has been awarded the American Chemical Society's Grady–Stack Award for interpreting chemistry to the public, and was the inaugural recipient of the Lagrange Prize for communicating complex science.

Brian Clegg read Natural Sciences, focusing on experimental physics, at the University of Cambridge. After developing hi-tech solutions for British Airways and working with creativity guru Edward de Bono, he formed a creative consultancy advising clients ranging from the BBC to the Met Office. He has written for *Nature*, the *Times*, and the *Wall Street Journal* and has lectured at Oxford and Cambridge universities and the Royal Institution. He is editor of the book review site www.popularscience.co.uk, and his own published titles include *A Brief History of Infinity* and *How to Build a Time Machine*.

Leon Clifford is a writer and a consultant whose speciality is simplifying complexity. Leon has a BSc in physics-with-astrophysics and is a member of the Association of British Science Writers. He worked for many years as a journalist covering science, technology and business issues with articles appearing in numerous publications including *Electronics Weekly*, *Wireless World*, *Computer Weekly*, *New Scientist* and the *Daily Telegraph*. Leon is interested in all aspects of physics – particularly climate science, astrophysics and particle physics.

Frank Close, OBE, is Professor of Physics at the University of Oxford and Fellow of Exeter College, Oxford. He was formerly Head of the Theoretical Physics Division, at Rutherford Appleton Laboratory and Head of Communications and Public Education at CERN. His research is into the quark and gluon structure of nuclear particles, where he has published more than 200 papers in the peer-reviewed literature. He is a Fellow of the American Physical Society, and of the British Institute of Physics, and won the society's Kelvin Medal in 1996 for his outstanding contributions to the public understanding of physics. He is the author of many books including *Neutrino* – short-listed for the Galileo Prize in 2013 – the best selling *Lucifer's Legacy: The Meaning of Asymmetry* and most recent *The Infinity Puzzle*.

Sophie Hebden is a freelance science writer based in Mansfield, UK. She combines writing about physics with looking after two small children. She has written for *New Scientist* and the Foundational Questions Institute, and is former news editor for SciDev.Net. She holds a PhD in space plasma physics, and a masters in science communication.

Alexander Hellemans is a science writer who has published articles in *Science, Nature, Scientific American*, BBC *Focus*, the *Guardian, New Scientist, The Scientist, IEEE Spectrum, Chemical and Engineering News* and other publications. With Bryan Bunch, Hellemans is the author of *The History of Science and Technology: A Browser's Guide to the Great Discoveries, Inventions and the People Who Made Them from the Dawn of Time to Today*. Previously both authors wrote *The Timetables of Science: A Chronology of the Most Important People and Events in the History of Science* and *The Timetables of Technology: A Chronology of the Most Important People and Events in the History of Technology*.

Sharon Ann Holgate is a freelance science writer and broadcaster with a doctorate in physics. She has written for newspapers and magazines including *New Scientist* and *Focus*, and presented programmes for BBC Radio 4, a mini-series for the BBC World Service and video interviews for the Myrovlytis Trust. She was co-author of *The Way Science Works*, a children's popular science book shortlisted for the 2003 Junior Prize in the Aventis Prizes for Science Books, and wrote the undergraduate textbook *Understanding Solid State Physics*. In 2006, Sharon Ann won Young Professional Physicist of the Year for her work communicating physics.

Andrew May is a technical consultant and freelance writer on subjects ranging from astronomy and quantum physics to defence analysis and military technology. After reading Natural Sciences at the University of Cambridge in the 1970s, he went on to gain a PhD in Astrophysics from the University of Manchester. Since then he has accumulated more than 30 years' worth of diverse experience in academia, the scientific civil service and private industry.

INDEX

ACKNOWLEDGEMENTS

PICTURE CREDITS

The publisher would like to thank the following individuals and organizations for their kind permission to reproduce the images in this book. Every effort has been made to acknowledge the pictures; however, we apologize if there are any unintentional omissions.

All images from Shutterstock, Inc./www.shutterstock.com and Clipart Images/www.clipart.com unless stated.

Bettmann/Corbis: 44, 62, 67.
German Federal Archives: 41.
H. Raab: 111.
Kelvin Fagan/Cavendish Laboratory: 128.
Keystone/Getty Images: 88.
Library of Congress, D.C.: 26, 99.
NASA: 59.
Queens University, Belfast: 102.
SSPL/Getty Images: 71.

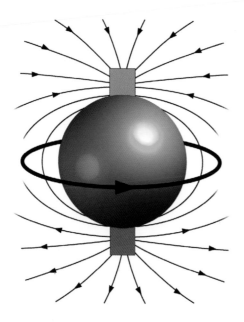